Industrial

Process

Sensors

Industrial

Process

Sensors

David M. Scott

Dupont Company
Experimental Station
Wilmington, Delaware, U.S.A.

CRC Press
Taylor & Francis Group
Boca Raton London New York

CRC Press is an imprint of the
Taylor & Francis Group, an **informa** business

CRC Press
Taylor & Francis Group
6000 Broken Sound Parkway NW, Suite 300
Boca Raton, FL 33487-2742

First issued in paperback 2019

© 2008 by Taylor & Francis Group, LLC
CRC Press is an imprint of Taylor & Francis Group, an Informa business

No claim to original U.S. Government works

ISBN-13: 978-1-4200-4416-4 (hbk)
ISBN-13: 978-0-367-38834-8 (pbk)

Library of Congress Cataloging-in-Publication Data

Scott, David M.
  Industrial process sensors / David M. Scott.
    p. cm.
  Includes bibliographical references and index.
  ISBN 978-1-4200-4416-4 (alk. paper)
  1. Engineering instruments. 2. Detectors. I. Title.

TA165.S38 2007
670.42'7--dc22   2007034223

**Visit the Taylor & Francis Web site at
http://www.taylorandfrancis.com**

**and the CRC Press Web site at
http://www.crcpress.com**

Dedication

To
L. A. Lundgren—teacher, mentor, and friend

Contents

Preface

Industry uses a variety of sensors to control its operations; the most familiar devices include thermocouples and pressure gauges, which measure a single variable at a single point in the process. As manufacturing processes have become more complex, additional types of information are required. Typical processes now need measurements of film thickness, particle size, solids concentration, and contamination detection. Most of these sensors operate on relatively simple principles that are based on the interaction between matter and sound, light, or electric fields. This book explains the physics upon which these measurements are based and provides additional details about sensor operation and interpretation of the data. Limits of sensitivity and significance are discussed, and examples are provided that illustrate how these devices have been used for process control to improve productivity or product uniformity.

There is often more than one type of sensor that will function adequately in a given application. For instance, the gauging (thickness measurement) of polyester film can be accomplished by using light, sound, or radiation. In such cases the choice of sensor always depends on the specific details of the application, so it is imperative to understand the operation and limitations of each device. Clearly, other factors such as cost or vendor issues need to be considered, but from a purely technical point of view the best choice of sensor for a given application ultimately depends on the details of the measurement process. The purpose of this book is to explore the relevant physics in order to facilitate that choice.

This monograph is based on 20 years of personal experience gained from working as a physicist in the chemical process industry, and the examples presented here are based on actual installations at nearly a dozen sites. Separate accounts of these examples have been published previously, but the aim here is to provide a coherent review of the physical mechanisms of process sensors. This book will be of interest to process owners, chemical engineers, electrical engineers, instrumentation developers, vendors, plant engineers, and plant operators from the chemical, mineral, food, and nuclear industries. I hope that this information will prove to be useful as they tackle even more challenging measurement needs.

David M. Scott
DuPont Company, Experimental Station,
Wilmington, Delaware, U.S.A.

Acknowledgments

This volume covers such a long time span that it is quite impossible to name all of the people who have contributed in some fashion to the work. It is certainly true that I owe a great deal to my colleagues at the Experimental Station, at the DuPont plants, and at various academic institutions here and abroad, but the list would be long indeed. I should, however, acknowledge the folks who collaborated with me to build and test the sensors described herein: Arthur Boxman, Chuck Fisher, Oliver Gutsche, John Harrington, Ed Jochen, Wouter Kuiper, Jerry Lee, Bob Moneta, Lou Rosen, Gregg Sunshine, and Robert Waterland. I am also grateful to Rajeev Gorowara, John Modla, and Christopher Scott for reviewing chapter drafts. Finally, I thank my wife, Sue, for her patience and support during the writing of this manuscript.

About the Author

David Scott is a physicist at the DuPont Company's main research facility in Wilmington, Delaware, where he has been developing industrial sensors and online measurement applications for two decades. He joined DuPont in 1986 after completing his Ph.D. in atomic and molecular physics at the College of William and Mary; he also holds the B.A. (Earlham College, 1981) and M.S. (William and Mary, 1984) degrees in physics. He initially worked on tomography and real-time radiography for nondestructive evaluation of advanced composite materials, and later developed optical sensors for industrial process applications.

In 1996 Dr. Scott was invited to establish a research group in the area of particle characterization; since that time, the scope of his group has expanded to include interfacial engineering and characterization of nanoparticle systems. His primary research interest is online characterization of particulate systems, and his research activities have included process imaging (including process tomography) and in-line ultrasonic measurement of particle size. He holds several patents, and has published over 30 technical papers in peer-reviewed journals, presented keynote and plenary lectures at many international conferences, and edited several special journal issues.

1

Introduction

1.1 Motivation for Process Measurement

Industrial companies exist to make money. The many societal benefits they provide (e.g., jobs, new technology, economic growth) are actually by-products of their primary function, which is to manufacture goods that can be sold at a profit in order to make money for their investors. Companies that in the long run continue to provide an adequate return on stockholder investment tend to survive; those that fail to do so eventually disappear. This somewhat simplistic view underscores the importance of profitable manufacturing operations, and it is ultimately the need to maximize profit that provides the motivation for a company to buy process measurement and control systems.

In this context, a process is a given set of manufacturing operations that converts raw material (such as chemicals or feedstock) into a useful product or an intermediate material that can subsequently be made into a useful product. In particular, the focus in this volume will be on measuring processes that involve physical or chemical changes in the state of the raw material, as opposed to simple machining operations. An example is the production of polymer film, which entails a chemical reaction at a suitable temperature and pressure, and the subsequent formation of the film. In this example the uniformity and quality of the product depend on the ability to maintain the correct temperature, pressure, and film thickness. Process sensors are devices that measure these and other parameters, and the resulting data is used to control the process. In addition, such measurements enable better process understanding, which often drives process improvement.

The connection between profit and process measurement is illustrated in Figure 1.1. Product quality and lot-to-lot uniformity generally have a large impact on market demand, especially if similar products are offered by competitors. Except in the case of a machine malfunction or operator error, defects are usually caused by variability in feedstock or excursions in operating conditions. It is possible to correct for some variability in the raw material by adjusting the process (e.g. changing the pH, or milling for a longer period of time). Feedback from process sensors enables active process control, which can respond to such needs as they arise. By automatically adapting the process to changing feedstock, the process controller improves product uniformity and minimizes the amount of defective product. The result is a top-quality product that will consistently meet customers' expectations.

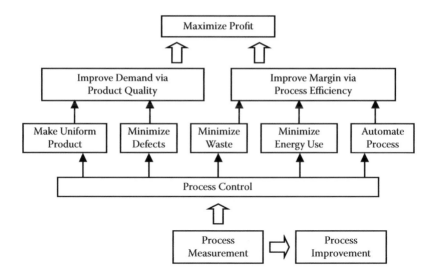

FIGURE 1.1 Process measurement is crucial to plant operation and profitability.

Process efficiency also has a major impact on profitability. By minimizing defective product, process control also minimizes the associated waste in raw material, effort, and energy. Automation of various operations within the process leads to lower labor costs. Finally, process measurement and control can reduce energy costs by running mills, mixers, and other energy-intensive devices only as long as necessary. It is not uncommon to realize savings of more than 15% in energy and maintenance costs on such equipment. The resulting increase in asset productivity is also important when the plant is running near capacity. By running an efficient process, the company can maximize its profit margin on the product.

Process measurement makes it possible to improve both product quality and process efficiency. Even without a closed-loop automatic control system in place, process sensors provide basic feedback to the operator. Since most processes operate entirely within closed metal vessels, the operator relies on sensor data for knowledge about the state of the process. It would therefore be impossible to run most plants without sensors.

1.2 Process Sensors

Modern plant operations are very complex and depend on a number of competing physiochemical mechanisms. For this reason, many unit operations require measurement capabilities far beyond traditional temperature and pressure readings. Table 1.1 lists a few types of process measurement along with the unit operations where they are needed and the industry segments in which such unit operations are used. A given measurement might be provided by more than one type of sensor, just as temperature can be measured with thermocouples, fluid-filled thermometers, and bimetal springs. How the sensors work is the topic of the subsequent chapters.

TABLE 1.1 Examples of Process Measurements

Measurement	Unit Operations	Industrial Segments
Concentration	Filtration Sedimentation	Chemical Pharmaceuticals
Droplet size	Emulsification	Agriculture
Film thickness	Casting Extrusion	Packaging Plastics
Flow rate	Material transport Reaction	Chemical Petrochemical
Fluid level	Filling	Food & Beverage Petrochemical
Homogeneity	Dispersion Mixing	Electronics Packaging
Particle size	Crystallization Granulation Milling	Agriculture Electronics Pharmaceuticals

Process sensors are categorized according to their placement. They are usually installed directly into a process vessel or pipe and provide a continuous readout. If the pipe is a side-stream or sampling loop, then the sensor is *online*; if, on the other hand, the bulk of the material being processed flows through or directly past the sensor, the sensor is an *in-line* or *in-process* one. Samples of material can also be extracted manually and fed to a nearby (*at-line*) instrument to determine the state of the process. In many cases samples are carried to another location, usually a quality control lab, for *offline* analysis of intermediate or product material. In process development applications, it is often expeditious or even necessary to use at-line or offline sensors. For this reason, a few examples of such measurements are included in later chapters.

Although an extensive array of process sensors is already available commercially, new sensors continue to be developed. Sometimes an industrial company develops a new sensor for a particular application in order to provide extended measurement range, faster response time, or new measurement capability. Since it can be expensive to develop new sensor technology, the most cost-effective solution is usually to adapt an existing sensor, if possible. Vendors continue to develop new sensors in order to attract more business or to stay ahead of their competition, but a new sensor is successful only if it provides novel measurement capability, an extended range of measurement, or lower cost compared to existing instruments. In any case, industry tends to invest in new technology only when the benefits are crystal clear.

The performance that industry wants from any process sensor, new or old, is summarized in Table 1.2:

1. The sensor must be completely reliable under continuous operation and ideally require no preventive maintenance. The sensor should be installed in such a way that it can be replaced quickly (in case it does eventually malfunction).
2. The sensor is easy to use and does not require a complicated calibration sequence.

TABLE 1.2 Summary of Expectations

What Industry Wants	What Industry Does Not Want
Reliability	Surprises
Ease of use	Nomadic technology
Relevant data	Complexity
Simple readout	Added expense
Compatibility	
Immediate payoff	

3. The data it provides is directly related to the physical properties of interest. A sensor that purports to measure viscosity but in reality is also sensitive to changes in pressure is of limited usefulness.
4. The data is easily interpreted. Sensors that measure scalar quantities such as temperature or flow rate generally output a signal that is proportional to these quantities. If the sensor interface is digital, the readout is provided in the correct units.
5. The sensor is compatible with other sensors and with the existing distributed control system.
6. The sensor must provide an immediate pay-off relative to its cost of purchase and installation (which is usually many times more than the purchase price).

There is a corresponding list of things that industry does not want (also listed in Table 1.2):

1. There should be no surprises in terms of cost, delivery, performance, or support.
2. Industrial plants are generally not interested in technology that is in search of new applications. The sensor must offer a clear solution to an actual manufacturing problem.
3. Complexity should be avoided because complicated hardware, software, and calibration procedures are difficult to support with a lean manufacturing staff. This caveat does not prohibit the technology behind advanced process sensors from being sophisticated, but the installation and use of the sensor must be perceived to be simple. A corollary is that a sensor should be no more complex than necessary.
4. The expense of a process sensor, including the installation cost, must be much lower than its perceived benefit.

1.3 The Physics of Measurement

All sensors rely upon interactions between matter and energy in order to make a measurement. At the most basic level, a process sensor detects changes in some measurable quantity (e.g. light intensity or electrical resistance) that occur in response to a physical change in state (e.g. temperature, pressure, or concentration) within the process. In order to understand how such sensors work, it is necessary to examine the physical mechanisms that they use.

What follows in this book is an exploration of how simple physical concepts have been applied to solve a variety of industrial measurement problems. The intent is not to dwell too much on the physics per se, but rather to examine the practical application of these ideas to process measurement (i.e., the physics of measurement). Since the operation of thermocouples and other simple process sensors is well understood, the focus here is on advanced sensors that meet difficult measurement problems.

The first part of this book is a review (in chapters 2–6) of basic concepts that are used in subsequent chapters. It is clearly impossible and counterproductive to recapitulate an entire undergraduate physics course in this volume, so only relevant highlights are covered; this is done in the hopes of reminding the reader of material already encountered in school. It is hoped that those to whom this presentation is entirely new will at least be able to understand the gist of it. Many related details have been glossed over in order to simplify the explanation and to save space, but the suggested reading lists identify textbooks that delve more deeply into the topics. Long mathematical derivations have been avoided or moved to an appendix where possible.

Chapter 2 introduces the generic sensor model, the requirements for a useful measurement, and the sensor output signal. Sources of noise and bias are discussed, and simple statistical tools are reviewed to enable a discussion of the propagation of errors in measurement; a more detailed example will be given in chapter 10.

The next chapters review basic physical phenomena. Chapter 3 reviews the phenomena associated with sound and other types of waves. The *wave equation* is derived in order to provide a mathematical description of wave shape and wave phenomena, which include reflection, refraction, diffraction, resonance, and Doppler shift. Several common devices for generating and detecting sound are described there. Chapter 4 discusses light and provides a brief survey of the operation of various optical components, including lenses, prisms, filters, lasers, and optical modulators. These devices affect the propagation of light and electromagnetic waves, and many of the components presented in this chapter are used in optical sensors. Chapter 5 introduces electricity, electrical measurements, and electronic components. Chapter 6 describes ionizing radiation, which includes electromagnetic waves of very short wavelength (e.g., X-rays), as well as highly energetic particles, such as subatomic particles produced by nuclear decay in radioisotopes.

Chapter 7 begins the applications section of the book and reviews conventional process sensors, including those that measure temperature, pressure, level, and flow rate. In the context of this volume, conventional sensors are those that meet three criteria: (1) they measure a single scalar quantity at a single point in the process; (2) they utilize technology that has been in widespread plant use for over two decades; and (3) they come from a variety of commercial sources. Many comprehensive books have already been written about the operation of such process sensors (see the suggested reading list at the end of each chapter), and so a complete survey need not be repeated here. The intent is simply to connect the physical phenomena described in the earlier chapters with some of the sensors in common use. It will also be shown that many quantities can be measured via multiple approaches.

Having addressed relatively simple measurement applications, the discussion will focus on more complicated measurement problems that could not be solved with conventional sensors. Case studies are provided in four broad areas. Chapter 8 introduces the concept of particle size distributions and describes how they can be measured in slurries, colloids and nanoparticle systems, and emulsions. Chapter 9 describes applications of imaging techniques, including real-time tomographic imaging of unit operations. Chapter 10 demonstrates the online measurement of film thickness for ultrathin films and the non-destructive measurement of individual ply thickness in multilayer films. The exposition is completed in chapter 11, which illustrates a variety of measurements in plastic and composite materials, including identification of polymer type for recycling operations, detection of contamination in molten polymer, characterization of particle dispersion in reinforced polymers and composites, and determination of part dimensions.

It is important to bear in mind that the presentation of these approaches does not imply a guarantee that they will be successful in other applications, nor does it convey any rights to practice the art or to infringe on any patented invention. In addition, since most problems have more than one solution, it should not be inferred that these approaches are necessarily the only ones that will work. These examples do, however, serve as existence theorems that prove such process measurements can be made, even under the demanding conditions found in most manufacturing operations. The hope is that these examples and explanations will shed some light on the solution to other, even more challenging industrial measurement problems.

2

Measurement

2.1 The Sensor Model

As depicted in Figure 2.1, a generic process sensor determines the current state of the process from one or more physical observables, which by definition are physical quantities that can be directly measured. Physical observables include temperature, frequency, light intensity, speed, and so forth. The sensor uses these inputs to produce a measurement, which may be a simple number (e.g., the local temperature) or a more complex data set such as an image.

If the measurement is to have any validity it must be true that a given set of physical observable values must always produce the same measurement value; otherwise, the measurement would be ambiguous. By taking a process measurement, we are assuming that the quantity to be measured *can* be measured and that it depends uniquely on the physical observables. This requirement is similar to the so-called zeroth law of thermodynamics (which states that temperature exists as a measurable quantity), and we could even call it the zeroth law of measurement. This general concept is described mathematically by a function S, whose domain is the set of all possible values of the physical observables and range is the set of all possible measurement values:

$$measurement = S(observables) \tag{2.1}$$

The fact that a sensor's output can be described by a function (at least in theory) follows directly from the causal relationships linking the quantity of interest in the process, the physical observable, and the resulting measurement.

Another requirement for making process measurements is that the function (equation 2.1) must be continuous, which suggests that infinitesimal changes in the physical observable can only cause infinitesimal changes in the measurement. This requirement ensures that the measurement data is differentiable and that the rate of change in the measurement remains finite.

A final requirement is that the measurement reported by the sensor must be linearly dependent on the process quantity to be measured. Ideally, the output of the sensor

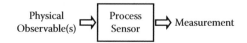

FIGURE 2.1 A generic process sensor.

(denoted as s) gives the exact value of the quantity (q) to be measured, but in practice there is often a small discrepancy given by

$$s = Aq + b \qquad (2.2)$$

where A is a gain factor and b is a constant offset. An important function of calibration is to adjust the sensor's electronics so that $A = 1$ and $b = 0$, thus ensuring that $s = q$. In most process control applications it is not necessary to get the calibration exact, because the controller usually bases its corrective actions on changes in the measurement rather than on the absolute value of the measurement.

The sensor model S represented by equation 2.1 is a set of equations that describe the physical mechanism of the interaction between the physical observable and the process. It is possible to formulate a mathematical model for any process sensor; whether or not the model captures the physics of the measurement is another matter. For instance, a proximity sensor that is based on electrical capacitance may be sensitive to changes in humidity or barometric pressure due to its design. If the model used by the sensor fails to include the environmental effects on the capacitance, then it will fail to describe the output of the sensor under all operating conditions. The sensor will always generate a signal, but the signal could be wrong. If however the model adequately describes the sensor design, then the sensitivity to these extraneous effects can be determined or even corrected.

2.2 Units of Measure

The sensor, together with whatever support circuitry is required, produces a measurement value that is usually represented by a voltage or current level, serial port signal, or other type of computer interface signal. The signal may be displayed on a panel if there is no direct computer connection. When measurements are represented by a voltage or current level (as in the case of 4–20 milliamp [mA] loop control systems), the pre-established range (minimum and maximum) of possible readings is used to interpret the sensor output. Thus a current output of 5 mA from one sensor might mean 1 meter per second (1 m/s), whereas 5 mA from a different sensor might mean 30°C. Therefore, not only the range but also the unit of measure is assumed. Sensors that communicate directly with a computer system generally do not have this ambiguity, since the units can be included in the string of characters that convey the measurement result.

The International System of Units include fundamental quantities such as the meter, kilogram, and second, and derived units such as the newton (the unit of force equal to 1 kg·m·s^{-2}). Since it is often necessary to express a reading that is either very large or very

TABLE 2.1 Prefixes Used with the International System of Units

Symbol	Prefix	Multiplier
P	peta	10^{15}
T	tera	10^{12}
G	giga	10^{9}
M	mega	10^{6}
k	kilo	10^{3}
c	centi	10^{-2}
m	milli	10^{-3}
μ	micro	10^{-6}
n	nano	10^{-9}
p	pico	10^{-12}
f	femto	10^{-15}

close to zero, the International System of Units also defines prefixes that multiply the base unit by a power of 10 (see Table 2.1).

2.3 Simple Statistics

Consider a new thermocouple that has been installed in the process and is now generating temperature data. Suppose the first reading is 30.0°C and the second reading is 30.2°C. Two possibilities can be considered: either the temperature is increasing or the thermocouple reading is uncertain by at least 0.2°C. All sensor readings are subject to a certain amount of variability, so in order to know whether or not the temperature is actually increasing, we must know whether or not a change of 0.2°C is significant in this case. In other words, if the temperature is truly constant, how much variability can be expected from the thermocouple readout?

Simple statistical descriptors can be used to assess whether or not a given variation in a reading is noteworthy. The first of these is the mean, or numerical average. Given a sequence of N readings $\{r_1, r_2, \ldots, r_N\}$, then the mean M is defined as

$$M = \frac{1}{N} \sum_{i=1}^{N} r_i \tag{2.3}$$

The standard deviation σ of the readings is defined to be

$$\sigma = \sqrt{\frac{1}{N} \sum_{i=1}^{N} (r_i - M)^2} \tag{2.4}$$

The significance of the standard deviation is that it provides an indication of the variability of the data. Small values of σ result from readings that are uniformly near the

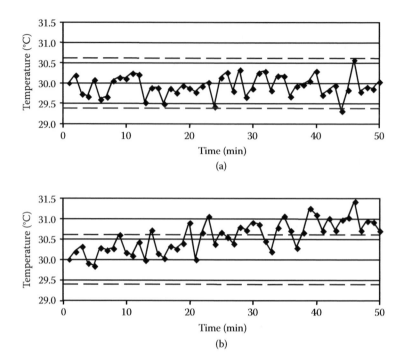

FIGURE 2.2 The output from a hypothetical thermocouple as a function of time when (a) the temperature is steady and (b) the temperature is increasing. The dashed lines indicate the plus and minus 2σ levels for the baseline reading.

mean, and large values of σ are caused by large variations from one reading to the next. The ratio (σ/M) is the coefficient of variation, which can be used to compare the variability between two sets of readings with different mean values.

As an example, consider the hypothetical thermocouple mentioned earlier, and suppose that its output is recorded once a minute for a period of 50 minutes, as shown in Figure 2.2a. The mean of these 50 measurements is 30.04°C, and the standard deviation is 0.30°C; therefore, the coefficient of variation is only 1%. Although there are random fluctuations from one reading to the next, there does not appear to be either an upward or a downward trend in the data. Since the standard deviation is 0.30°C and the difference between the first two readings is only 0.20°C, it seems likely that the initial increase in the reading was not significant (i.e., no action should be taken by the process controller).

The measurement values in this hypothetical example are normally distributed, which is to say that the probability P that the reading equals r is given by

$$P(r) = \frac{1}{\sigma\sqrt{2\pi}} \exp\left[-\frac{(r-M)^2}{2\sigma^2}\right]$$

(2.5)

where the mean M and standard deviation σ of the measurement are defined above. The difference $(r - M)$ is usually interpreted as instrumental error. It is known from statistical analysis that in the special case of normally distributed data, about 68% of the values are within one standard deviation of the mean, and about 95% are within two standard deviations.[1] Only one in a million readings will deviate farther than 5σ from the mean (assuming that the physical quantity being measured actually remains constant). In this example $\sigma = 0.3°C$, so 95% of the readings are expected to lie in the range of 29.4°C to 30.6°C as indicated by the dashed lines in Figure 2.2a. In fact, the figure shows that 49 of the 50 readings (i.e., 95%) do lie within this region, which means that the temperature is stable according to this hypothetical thermocouple. If it is known that the standard deviation is 0.3°C, then it becomes clear that the apparent increase of 0.2°C in the second temperature reading is not a significant change.

Figure 2.2b shows the output of the same hypothetical thermocouple during a heating cycle, with the plus and minus 2σ limits indicated by dashed lines as before. The first two readings just happen to be 30.0°C and 30.2°C, as before. As argued above, the increase in the second reading is not large enough to indicate that the temperature is increasing. However, after about 20 minutes the temperature readings frequently exceed the limit indicated by the upper dashed line. At that point it is clear that the temperature has increased by an amount that is statistically significant.

The preceding example demonstrates that a discussion of the reproducibility of a measurement is closely tied to a consideration of its statistics. In particular, the expected variability in a measurement can be quantified by the standard deviation. Although we have not considered the root cause of the variability in the measurement, it is clear that this variability must be reduced in order to improve the precision of the measurement.

2.4 Sources of Error

The variability observed in multiple sensor readings is a combination of the real variability in the quantity being measured and the variability due to the sensor itself. The latter is usually called the *instrumental error*. The goal of the measurement is to determine the process parameters (temperature, flow rate, etc.), and any variability in the process should become evident from a statistical review of the process sensor data. As demonstrated above, the interpretation of this data depends on the standard deviation of the sensor output under quiescent conditions. Therefore, the instrumental error must be assessed in order to determine the significance of changes in the sensor output.

The sources of instrumental error can be classified into *systematic error* and *stochastic (random) error*. Systematic error is usually due to a flaw in the design, installation, calibration, or use of the process sensor. This type of error is manifested as a constant, or slowly changing, offset or multiplicative factor in the measurement. Common causes of systematic error are sliding mechanical seals in force or position sensors, dark current in infrared detectors and video cameras, and incorrect values of physical constants (such as viscosity or index of refraction) in calculations. The accumulation of process material on the windows of optical sensors ("fouling") will also lead to systematic errors as a result of the decrease in light intensity.

Systematic errors often lead to instrumental drift, wherein the amount of error changes relatively slowly over time. Since electronic circuits are sometimes sensitive to changes in temperature, a potential source of instrumental drift is the ambient temperature in the process area. Depending on the operational design of the sensor, other environmental conditions (such as barometric pressure) can contribute to drift.

Stochastic errors are rapid but usually small fluctuations in the sensor output, and this random component of the signal is often called *noise*. Sensor noise comes from a variety of sources including vibrations from the process equipment, ambient sound, electrical ground loops, corroded electrical connections, and electronic circuitry. Mechanical vibrations can interfere with the precise optical or mechanical alignment in a sensor and cause an error in measurement. Sensors that are designed to measure very small electrical signals contain amplifier circuits with a high gain; such devices are often sensitive to vibration. Sensors that are very sensitive to vibration are microphonic and therefore respond to ambient sound in the process area (often a very noisy environment).

Ground loops are formed by connecting electrical equipment to ground via multiple paths or ground circuits. These connections create a continuous circuit or loop in which alternating currents are induced by neighboring electrical equipment. The induced current leads to a fluctuating voltage offset as a result of the resistance of the wiring, and this offset appears as a spurious signal in the sensor's electronics.[2] In a process environment, the current flowing through ground loops can be quite strong, so the process sensors must be adequately shielded and properly grounded in order to provide reliable results.

Mechanical vibrations and electrical ground loops contribute stochastic noise to the process measurement, but it often happens that such noise is dominated by a few component frequencies. Truly random electrical noise is caused by loose or corroded electrical connections and electronic amplification. When an electrical connection becomes loose or corroded, a slight voltage difference may develop across the junction. This voltage is not well defined because the junction resistance is not well defined, and the fluctuating voltage affects the sensor output in an unpredictable manner. Other sources of random electrical noise include *shot noise* and *thermal noise* (see for example Fraden 1996, pp. 212ff). Shot noise is caused by the fact that electrical current is conducted by electrons; the current appears to be continuous, but it is in reality a rapid and random succession of charge transfer events. Thermal noise is due to the random motion of electrons in a conductor; this motion is dependent on the temperature of the conductor, so an increase in temperature produces an increase in thermal noise.

Noise can often be mitigated by averaging the sensor's output signal over a number of readings. Stochastic error is as likely to cause an increase in signal as it is a decrease, so the positive and negative fluctuations can be expected to cancel each other if a sufficiently large number of measurements are summed. An example is provided in Figure 2.3, which shows a 10-point moving average of the data shown in Figure 2.2.[3] The constant temperature case (Figure 2.3a) shows a fairly constant output, and the heating cycle case (Figure 2.3b) clearly indicates an increase in temperature. Process sensors often perform some internal signal averaging in order to produce a more stable signal.

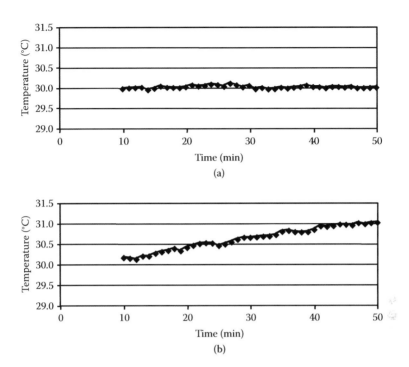

FIGURE 2.3 A 10-point moving average of the data shown in Figure 2.2 for a hypothetical thermocouple: (a) the moving average of data shown in Figure 2.2a; (b) the moving average of data shown in Figure 2.2b.

2.5 Analysis of Error

The individual systematic and stochastic errors are involved in making a measurement propagate through the process sensor and become compounded in the resulting output signal. If the sensor model is known, it is possible to estimate the relative effect of each of these contributions through an analysis of error. After one has identified the relative contributions of each source of error, it is possible to focus on reducing the most egregious of them. A thorough example of error analysis is given in chapter 10, where the effects of environmental conditions on a film thickness sensor are considered.

Error analysis is essentially a calculation of the cumulative effect caused by either small changes or systematic error; it is therefore a calculation of differentials. A simple example is the measurement of resistance, which is the basis of many simple sensors. If a constant current I is passed through a resistance R, the voltage difference V between its two terminals (see chapter 5) is given by

$$V = IR \tag{2.6}$$

Therefore, knowing the amount of current, one could measure the voltage and calculate the resistance according to

$$R = \frac{V}{I}$$

(2.7)

Thus, two distinct quantities are involved in this measurement, and the total error in measured resistance depends on the error in measured voltage and the error in the assumed current.

The absolute error (δR) in the measured resistance due to the error (δV) in the measured voltage is

$$\delta R\big|_I = \frac{\partial R}{\partial V}\delta V = \left(\frac{1}{I}\right)\delta V$$

(2.8)

where ∂ denotes partial differentiation and the vertical bar signifies that the current is held constant. The relative or fractional error ($\delta R/R$) can be calculated by dividing both sides of equation 2.8 by equation 2.7:

$$\frac{\delta R}{R}\bigg|_I = \left(\frac{I}{I}\right)\frac{\delta V}{V} = \frac{\delta V}{V}$$

(2.9)

Similarly, it can be shown that the fractional error due to an error in the assumed value of the current (which may have changed over time) is given by

$$\frac{\delta R}{R}\bigg|_V = -\frac{\delta I}{I}$$

(2.10)

The negative sign indicates that the resistance is underestimated if the current is overestimated.

Since the measurement of voltage is independent of the estimation (or prior measurement) of the current, the contributions to the fractional error are also independent and must be added in quadrature. Therefore, the total fractional error in the measured resistance is

$$\frac{\delta R}{R} = \sqrt{\left(\frac{\delta R}{R}\bigg|_I\right)^2 + \left(\frac{\delta R}{R}\bigg|_V\right)^2} = \sqrt{\left(\frac{\delta V}{V}\right)^2 + \left(\frac{\delta I}{I}\right)^2}$$

(2.11)

Equation 2.11 shows that the relative error of the resistance measurement is greater than the larger of the two relative errors in voltage and current. If those two error terms are comparable to each other, then

$$\frac{\delta R}{R} \approx 1.414\left(\frac{\delta V}{V}\right)$$

(2.12)

Results similar to equations 2.11 and 2.12 can be derived for any process sensor if the function S from equation 2.1 is defined. The value of such calculations is that they provide a rigorous framework for understanding sensor error in terms of the error in each of the physical observables involved in the measurement. Once the sensor variability is defined, it is possible to address the variability of the process itself.

Suggested Reading

Bevington, P.R. and Robinson, D.K. (1992). *Data Reduction and Error Analysis for the Physical Sciences,* Second Edition. New York: McGraw-Hill.

Triola, M.F. (2005). *Elementary Statistics,* Ninth Edition. Boston: Pearson/Addison-Wesley.

3

Sound and Wave Phenomena

3.1 Sound

Hearing is one of the five basic senses, and sound therefore has always been a means of gathering information about one's immediate environment. For instance, an explorer who is looking for a stream naturally follows the sound of running water. Anyone who works with wood knows that it is usually possible to determine the quality of a board by tapping on it and listening to the sound. It should therefore come as no surprise that industrial sensors use sound to measure a variety of parameters such as distance, film thickness, particle size, and solids concentration. These applications will be discussed in later chapters; the purpose of this chapter is to review the properties of sound and to introduce concepts that will be used later.

What we perceive as sound is a set of momentary fluctuations in local air pressure caused by a mechanical motion, such as the ringing of a bell (Figure 3.1). As the surface of the bell moves imperceptibly and rapidly, it pushes on the molecules in the air and sets a wave of alternating compression and rarefaction in motion. A wave is a disturbance that moves through a given medium or field, and in the case of Figure 3.1 the regions of compression (called the *wave front*) and rarefaction move to the right. The sound is heard when this wave exerts a varying pressure on the eardrum. These vibrations are carried to the inner ear, where they are sensed within the cochlea.

In air, sound propagates as a longitudinal wave, which means that the air molecules oscillate back and forth along the direction of wave propagation. Figure 3.2 depicts the gas molecules in a hypothetical sound wave at five different moments in time as they move along the direction of propagation. The regions where the molecules are bunched together is a local high-pressure region, and the graph above each cartoon of the air molecules shows the local pressure $P(x)$ as a function of position x. The pressure gradient drives the molecules forward at high velocity, and when these molecules impact their neighbors, they impart momentum and recoil. Thus, a given molecule does not travel far before it transfers the momentum of the wave. After 1.47 milliseconds (ms), the local maxima have moved to the right a distance of 0.5 meters (m). At 3.94 ms the maxima have moved a total of 1.0 m, which happens to be the wavelength as defined below. It should be

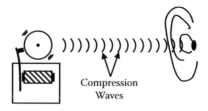

FIGURE 3.1 The ringing of a mechanical bell creates sound waves.

noted that, neglecting self-diffusion, the air molecules end up back at their original positions even though the wave itself has moved forward by one complete cycle. Of course, sound tends to spread out in all directions like ripples in a pond, but under certain conditions a directional beam of sound can be generated (see section 3.4.6).

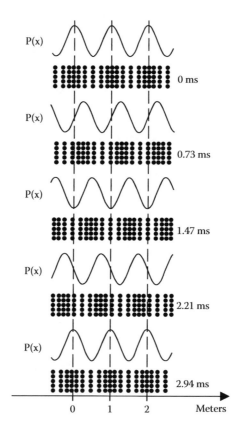

FIGURE 3.2 The propagation of pressure waves through air, shown at five different instants in time. The dots represent molecules of air, which move forward and backward along the direction of propagation. The local compression of air molecules causes variations in the pressure *P(x)*, which is shown as a function of position *x*.

FIGURE 3.3 Two types of wave: (a) longitudinal; (b) transverse.

Sound is transmitted by solids, liquids, and gas. In solid materials, sound propagates not only as a longitudinal (compression) wave but also as a transverse (shear) wave. These two types of wave travel at different speeds, the shear wave traveling at approximately half the speed of the compression wave. Shear waves do not travel very far through fluids (air or water) because there is no long-range order. Another example of a transverse wave is the motion of a guitar string that has been plucked; each element of the string moves in a direction that is transverse to the motion of the wave. As in the case of longitudinal waves, the elements of the string do not move very far from their original starting point (i.e., it is the wave that propagates, not the medium). The difference between longitudinal and transverse waves is depicted in Figure 3.3 (note that these waves do not have the same wavelength).

3.2 Waves

It turns out that many physical phenomena besides sound can be described by a wave. In general, there are three distinct classes of wave. *Mechanical waves* comprise the most familiar class, which includes sound, earthquakes, ocean waves, and the motion of a guitar string. *Electromagnetic waves* (see chapter 4) include X-rays, light, and radio waves. *Matter waves* are used in quantum mechanics to describe atoms and subatomic particles such as electrons and neutrons. A discussion of matter waves is outside the scope of this book, but it is noted in passing that phenomena associated with waves (such as interference, which is described in section 3.4.3) have also been observed in the case of electrons and neutrons.

The *frequency* of a wave refers to the rapidity with which the medium (in this case, air) oscillates. The unit of frequency is the *hertz* (defined to be one cycle per second), which is abbreviated as Hz.[1] A continuous wave at a single frequency is shown in Figure 3.4. The top part of Figure 3.4a shows the time evolution of what happens at a fixed position when the wave passes by it. As a specific example, the y-axis could represent the elevation of a cork in a pond as it rides over ripples on the surface of the water. The bottom part of Figure 3.4b shows a picture of the spatial extent of the wave, recorded at a single instant in time. Both waveforms are identical in shape; the time it takes to complete a full cycle of the wave (i.e., the duration) is called the *period* (as shown in Figure 3.4a), and the corresponding spatial quantity (shown in Figure 3.4b) is the wavelength. The frequency f, in hertz, is determined from the period T:

$$f = \frac{1}{T} \tag{3.1}$$

The frequency range of audible sound extends from about 20 Hz to 20 kHz, depending on the age of the hearer. Lower frequencies are felt rather than heard, and higher

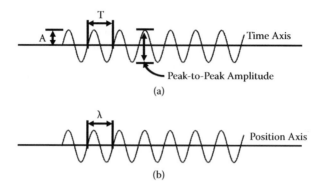

FIGURE 3.4 A continuous wave at a single frequency: (a) the time dependence of a wave measured at a fixed point; the amplitude A and the peak-to-peak amplitudes are shown; the period T is the time interval between zero crossings at the axis (or equivalently, the interval between any two successive points of the wave that have the same phase); (b) The position dependence of a wave at a single instant in time; the wavelength λ is the distance between zero crossings at the axis (or equivalently, the distance between any two successive points of the wave that have the same phase).

frequencies are inaudible; sound at frequencies beyond human hearing is called *ultrasound*. In practical applications ultrasound is limited to the range of 20 kHz to about 100 MHz because at higher frequencies (approaching 1 GHz) most materials heavily attenuate the wave. Ultrasound does not propagate very far in air above 1 MHz because the air molecules do not interact fast enough to transmit the sound. For comparison, the frequency of visible light is roughly half a billion times higher, ranging from 430 to 750 THz.

Waves can take on a variety of shapes, as depicted in Figure 3.5. The simplest waveform is a continuous wave (CW) of infinite extent (shown in Figure 3.5a), which represents a single frequency or tone. A tone burst is a slightly more complex waveform in which a single frequency is turned on for a short duration (typically five cycles or so). Figure 3.5b shows a double tone burst at two frequencies; the second burst is at a higher

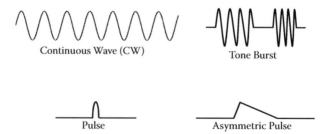

FIGURE 3.5 Four varieties of wave shape, as discussed in the text.

frequency than the first. A simple pulse is shown in Figure 3.5c, and an asymmetric pulse in Figure 3.5d.

In suspensions (solid particles suspended in liquid) and solid mixtures (e.g., concrete), the longitudinal wave is dissipated (attenuated) when it scatters from the particles; measurements based on this effect are discussed in chapter 8. Even in the absence of particles to scatter the sound, both solids and fluids have an intrinsic attenuation that reduces the sound intensity over distance.

3.3 The Wave Equation and Its Solutions

It can be shown (see section 3.6) that a sound wave $\psi(x,t)$ is a solution of the wave equation:

$$\frac{\partial^2 \psi}{\partial x^2} = \frac{1}{c^2}\left(\frac{\partial^2 \psi}{\partial t^2}\right) \tag{3.2}$$

where x is position, t is time, and c is the speed of sound. Solutions to equation 3.2 are periodic functions and linear combinations of periodic functions. The sine function is one of the most familiar solutions to the wave equation; the displacement at a fixed position for the wave in Figure 3.4a can be represented by the function $A\sin(\omega t)$, and the wave in Figure 3.4b can be described by $A\sin(kx)$. In general a continuous wave ψ, which is a function of both position and time, can be represented by either of these two equations:

$$\psi(x,t) = A\sin(kx - \omega t + \varphi) \tag{3.3a}$$

$$\psi(x,t) = A\sin(kx + \omega t + \varphi) \tag{3.3b}$$

where (at least for now) A, k, ω, and φ are constants. It is easily demonstrated by substitution that equations 3.3a and 3.3b are solutions of equation 3.2 if and only if

$$c = \frac{\omega}{k} \tag{3.4}$$

The quantity c is the velocity of the wave, which is a material property of the medium in which the wave travels. Equation 3.3a represents a wave traveling to the right, and equation 3.3b represents a wave traveling to the left.

The constants in equation 3.3 determine the size, wavelength, speed, and phase of the wave. The amplitude of the wave is given by A, so the peak-to-peak amplitude equals $2A$ as shown in Figure 3.4a. The phase offset φ advances or retards the wave by a fixed phase angle. It has already been noted in equation 3.1 that the frequency of the wave is the inverse of its period; the angular frequency, usually denoted as ω, is the frequency expressed in units of radians per second. Since an angle of 2π radians equals 360 degrees, it follows that frequency f, the period T, and angular frequency ω are related by

$$\omega = 2\pi f = \frac{2\pi}{T} \tag{3.5}$$

In a similar manner, the wave number k is inversely proportional to the wavelength λ:

$$k = \frac{2\pi}{\lambda} \tag{3.6}$$

The frequency and wave number are related through equation 3.4, which implies that the wave speed is the product of wavelength and frequency:

$$c = \lambda f \tag{3.7}$$

Since the speed of sound in dry air at standard temperature and pressure is about 340 m/s, the sound depicted earlier in Figure 3.2 (which has a wavelength of 1 m) has a frequency of 340 Hz. This pitch corresponds approximately to the musical note F above middle C.

A still more general solution of the wave equation is the plane wave, given by

$$\psi(x,t) = e^{i(kx - \omega t + \varphi)} \tag{3.8}$$

where

$$e^{i\theta} = \cos\theta + i\sin\theta \tag{3.9}$$

It is known that a *Fourier transform* can be used to express an arbitrary function $\psi(t)$ as an integral over all possible frequencies:

$$\psi(t) = \frac{1}{\sqrt{2\pi}} \int_{-\infty}^{\infty} A(\omega) e^{i\varphi(\omega)} e^{i\omega t} d\omega \tag{3.10}$$

where A and φ, both real functions of the angular frequency ω, are the amplitude and phase angle of a complex number given by

$$A(\omega) e^{i\varphi(\omega)} = \frac{1}{\sqrt{2\pi}} \int_{-\infty}^{\infty} \psi(t) e^{-i\omega t} d\omega \tag{3.11}$$

It should be evident that the wave shapes shown in Figure 3.5 can be transformed into Fourier components using equation 3.11; since each of the components is a solution of the wave equation, it follows that these other wave shapes are also solutions of the wave equation.

Until this point the assumption has been made that the wave speed is a constant, but it turns out that in physical systems the wave speed c is a function of frequency. Sometimes the change in wave speed is quite small over the frequency range of interest, in which case it can be neglected. When the speed changes significantly as function of frequency, it is necessary to distinguish between the phase velocity c (given by equation 3.4) and the group velocity c_g, which is given

$$c_g = \frac{\partial \omega}{\partial k} \tag{3.12}$$

The distinction is that the individual Fourier components of the wave travel at their individual phase velocities, whereas the wave itself travels at the group velocity. If the individual components travel at different speeds, then over time (or distance) the phase $\varphi(\omega)$ between the components is no longer correct, and the shape of the wave becomes distorted. This effect is called *dispersion*, which is not to be confused with the dispersion of small particles in a liquid (discussed in chapter 8).

3.4 Wave Phenomena

3.4.1 Reflection

When a sound wave strikes a wall, it bounces back as an echo. When water waves hit the side of a pool, they also reverse direction. Both of these examples demonstrate that waves can be reflected at the interface between two different media. The propagation of sound in a medium is governed by the acoustic impedance z, defined to be

$$z = \rho c \tag{3.13}$$

where ρ is the density and c is the speed of sound. When there is a discontinuity in the acoustic impedance—for instance, at the surface of a wall—a portion of the sound is reflected and a portion is transmitted. A similar effect occurs when light strikes the interface between transparent materials of different refractive index n (as shown in Figure 3.6a) and in the case of electrical signals that encounter a change in electrical impedance. It should be noted that the angle of reflection is equal to the angle of incidence.

3.4.2 Refraction

Figure 3.6a also shows that some of the wave can be transmitted at the interface. Since the propagation speed is generally different in the two media, the wavelength changes

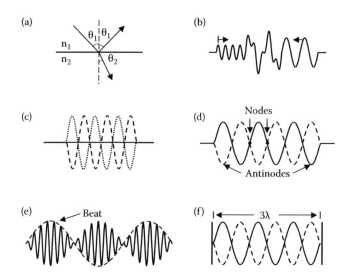

FIGURE 3.6 Generic wave phenomena: (a) reflection and refraction; (b) the superposition of two colliding waves; (c) the interference of two waves (the dotted and dashed lines) causes cancellation (solid line) at this moment in time; (d) a standing wave; (e) the beat signal caused by combining two signals of different frequencies (as in heterodyning); (f) the resonance of a cavity.

at the interface. The speed of the transmitted wave c_2 is related to the speed c_1 of the incident wave via

$$c_2 = \left(\frac{n_1}{n_2}\right)c_1 \tag{3.14}$$

where n_1 and n_2 are the refractive indices of the two materials. It follows from equation 3.7 that the wavelengths are related by

$$\lambda_2 = \left(\frac{n_2}{n_1}\right)\lambda_1 \tag{3.15}$$

In the case of normal incidence, which means that θ_1 equals zero, the wave continues to propagate in the same direction. At nonzero angles of incidence, however, the mismatch in wave speed causes the transmitted beam to propagate in a different direction. The change in direction is known as refraction, and the angle of refraction θ_2 is related to the angle of incidence θ_1 via Snell's Law:

$$\sin\theta_2 = \frac{n_1}{n_2}\sin\theta_1 \tag{3.16}$$

Since the index of refraction n and the acoustic impedance z are in general a function of frequency, the angle of refraction also depends on the frequency of the wave. For this reason a beam of white light is split into its component colors when it passes through a prism.

It follows from equation 3.16 that when $n_1 > n_2$ a critical angle θ_c exists at which $\theta_2 = 90°$, and above which none of the wave is transmitted. This critical angle is given by

$$\theta_c = \arcsin\left(\frac{n_2}{n_1}\right) \tag{3.17}$$

3.4.3 Superposition and Interference

Whenever two waves encounter each other as they travel through the medium, the net displacement of the medium at any point in space or time is simply the sum of the displacements of the individual waves. This effect is called *superposition*, and it is illustrated in Figure 3.6b for the case of two generic waves traveling in opposite directions. Each wave continues to move at a constant speed, and eventually the waves pass through each other without distortion (assuming the amplitudes are not large enough to cause nonlinearity in the medium). In general, for two waves ψ_1 and ψ_2 (which are both functions of position and time), the combined wave ψ is given by

$$\psi = \psi_1 + \psi_2 \tag{3.18}$$

The superposition of two sinusoidal waves leads to the phenomenon of interference. When two identical waves are a half-cycle (π radians) out of phase they cancel each other exactly; the result is destructive interference, as depicted in Figure 3.6c. From the trigonometric identity $\sin(\alpha + \pi) = -\sin(\alpha)$ it follows that

$$\sin(kx - \omega t) + \sin(kx - \omega t + \pi) = \sin(kx - \omega t) + [-\sin(kx - \omega t)] = 0 \tag{3.19}$$

Constructive interference occurs when the waves are in phase with each other; the result is a sinusoid with twice the amplitude of the original waves. At intermediate values of phase, the two waves combine to yield a sinusoid of the same frequency but with an amplitude that depends only on the phase difference φ. It is easily shown that

$$\sin(kx - \omega t) + \sin(kx - \omega t + \varphi) = 2\cos\left(\frac{\varphi}{2}\right)\sin\left(kx + \omega t + \frac{\varphi}{2}\right) \tag{3.20}$$

An acoustic example of interference is depicted in Figure 3.7, which shows sound waves emanating from two identical sources, A and B. Destructive interference creates a spatial pattern of dead zones where the sound intensity is significantly reduced, and this pattern is determined by the spacing between A and B and the frequency of the signals.

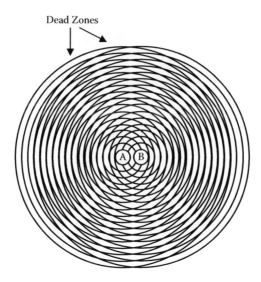

FIGURE 3.7 Two speakers that emit the same tone create an interference pattern that includes dead zones in which the sound is reduced.

For this reason the placement of the speakers in a stereo system has an impact on the quality of the sound produced.

A special case of superposition occurs when two sinusoidal waves with the same amplitude and frequency travel in opposite directions. As the two waves pass through each other, the relative phase between them changes linearly with time. Due to constructive and destructive interference, the net result is that the displacement at every point of the wave alternates between zero and some maximum value (Figure 3.6d). The result is a *standing wave*, which is so named because it does not appear to move but simply oscillates in place. Standing waves have equally spaced locations of maximum lateral displacement (called *antinodes*) alternating with locations of zero displacement (*nodes*), which never move.

It can also be shown that at any fixed position the superposition of two sine waves with different frequencies ω_1 and ω_2 results in an amplitude modulated signal (Figure 3.6e) whose carrier frequency is the average of ω_1 and ω_3. The modulation occurs at a frequency equal to half the difference in frequencies, so that

$$\sin(\omega_1 t) + \sin(\omega_2 t) = 2\sin\left[\left(\frac{\omega_1 - \omega_2}{2}\right)t\right]\sin\left[\left(\frac{\omega_1 + \omega_2}{2}\right)t\right] \tag{3.21}$$

The quantity $(\omega_1 - \omega_2)/2$ is known as the "beat" frequency, and it goes to zero as ω_1 approaches the value of ω_2. When the two frequencies differ by only a few hertz this beat can easily be heard. This phenomenon is well known to musicians, who use it to tune their instruments. The process of combining two signals to produce a lower frequency is called *heterodyning*, and it is widely used in radio circuits.

3.4.4 Resonance

It has been shown above that waves can be reflected, and that two identical waves traveling in opposite directions create a standing wave. If two reflectors are positioned so that a reflected wave can travel back and forth between them, a resonator is formed (Figure 3.6f). Whenever the frequency of the wave is such that the distance between the reflectors is an integral number of wavelengths, a standing wave is created. In this resonant condition the amplitude of the wave is at its maximum, so the sound is loudest (or the light is brightest). Dissipation of the acoustic or optical energy causes the wave to decrease over time, so a source of energy is needed to maintain the wave indefinitely. Wind instruments are familiar examples of acoustic resonators.

3.4.5 Doppler Shift

When a stationary source of sound emits a wave at a fixed frequency, a stationary observer hears or sees the wave at that same frequency. However, if the source and observer are moving relative to each other, the observed frequency f_d differs from the emitted frequency f_s. To see why this is so, consider Figure 3.8, which shows an acoustic source (A) moving in still air toward an observer (C) at a speed v_s. The period of the wave is the time T between successive wave fronts (Figure 3.4a). If the speed of sound is denoted as c, then during one period a given wavefront moves a distance cT while the source moves a distance v_sT. Therefore the observed distance between successive wave fronts is ($cT - v_sT$), which is by definition the observed wavelength. The frequency observed at position C is found from equation 3.7 to be

$$f_d = \frac{c}{\lambda_d} = \frac{c}{(cT - v_sT)} = \left(\frac{c}{c - v_s}\right)\left(\frac{1}{T}\right) = \left(\frac{c}{c - v_s}\right)f_s \qquad (3.22)$$

Thus, the tone heard by observer C is at a higher frequency than the emitted tone.

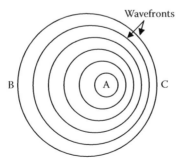

FIGURE 3.8 A Doppler shift caused by motion of the source A moving away from observer B and toward observer C.

This change in frequency is known as the *Doppler shift*, and the phenomenon itself is called the *Doppler effect*. If the detector (observer C) is also moving toward the source at a velocity v_d then

$$f_d = \left(\frac{c+v_d}{c-v_s} \right) f_s \tag{3.23}$$

As soon as the source passes the observer, the observed frequency of the receding source decreases to

$$f_d' = \left(\frac{c-v_d}{c+v_s} \right) f_s \tag{3.24}$$

giving the familiar drop in pitch heard if a car horn honks as it goes by. When source and observer approach one another, the Doppler shift is toward a higher frequency; when they draw apart the shift is toward a lower frequency. During this entire time, stationary observer B in Figure 3.8 hears only a lower frequency, because the source is moving away.

A similar effect is observed in the case of light, where the longitudinal Doppler shift (see Halliday et al., 2005, p. 1039) is given by

$$f_d = \left(\sqrt{\frac{1-\beta}{1+\beta}} \right) f_s \tag{3.25}$$

Here the relativistic factor $\beta = v/c$, c is the speed of light, and v is the relative velocity between source and observer. In equation 3.25 the source and detector are separating, and the result is a "red shift" to a lower frequency; if the source and detector are approaching each other, the signs in equation 3.25 are switched and the result is a "blue shift" toward a higher frequency. At low speeds where $\beta \ll 1$, equation 3.25 can be expanded in terms of β; to first order it becomes

$$f_d = (1-\beta)f_s = \left(\frac{c-v}{c} \right) f_s \tag{3.26}$$

which is equivalent to the result in equation 3.23 obtained in the case of sound waves for a stationary observer.

In contradistinction to the case of sound waves, the Doppler effect in light occurs even if the source and detector are not moving in the direction of the light beam. The *transverse Doppler shift* is due to the relativistic effect of time dilation that occurs at high speeds (see for example Halliday et al., 2005, p. 1041):

$$f_d = \sqrt{1-\beta^2}\, f_s \tag{3.27}$$

3.4.6 Diffraction

As shown in Figure 3.9, when a plane wave passes through an aperture its wavefront becomes distorted, and the wave spreads out. Interference between different phases of the wave creates a spatial pattern of varying intensity. This effect is called *diffraction*, and it has been observed to occur for sound, light, and subatomic particles.

Diffraction may be explained by considering the incident plane wave as an ensemble of individual wavelets, each with the same wavelength and phase relative to the other wavelets. These wavelets are called *Huygens waves*, after the seventeenth-century Dutch scientist Christiaan Huygens, who introduced them to describe the refraction of light. These wavelets expand in all directions and join with neighboring wavelets to form the next wavefront (see Figure 3.10). Due to symmetry, the portions of the wavelets that are transverse to the direction of propagation cancel each other. As long as there are no obstructions of changes in wave speed, the wave continues to propagate in the same direction.

Referring back to Figure 3.9, the incident wave can be regarded as composed of Huygens waves. As the wave passes through the aperture, most of the wavelets that comprise the original wave are blocked by the barrier. The wavelets that are not blocked continue to propagate. In essence, each point in the plane of the aperture acts as a tiny wave source that is in phase with all the others. As these waves spread out (to the right in this figure), they interfere with each other. In order to keep track of phase, in this figure the solid lines represent the crests (wave fronts) of the waves, and the dashed lines represent the troughs; many of the Huygens waves have been omitted for clarity. It can be seen that, in the region along the original direction of propagation, the wavelets remain more or less in phase. However, the wavelets become out of phase with increasing distance from the centerline of the aperture, and eventually they cancel each other. This cancellation creates a nodal line similar to the dead zones seen in Figure 3.7, which shows the nodal structure created by two speakers.

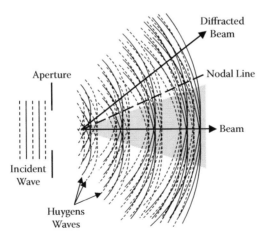

FIGURE 3.9 Diffraction of a wave by an aperture. Wave fronts are represented by solid lines, and troughs are represented by dashed lines. Many of the Huygens waves have been omitted for clarity.

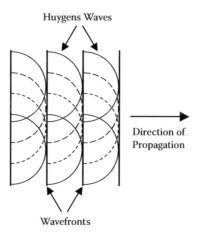

Huygens Waves

Direction of
Propagation

Wavefronts

FIGURE 3.10 Reconstruction of wavefronts by Huygens waves.

Beyond this line there is another region where the cancellation is not complete, and the wavelets there combine to form the diffracted beam. Depending on the ratio of the aperture diameter to the wavelength, it is possible to create more than one diffracted beam.

The wave in Figure 3.9 propagates forward from the aperture as a widening beam, which is depicted as a gray wedge. The relatively weak diffracted beam exits at an angle, and there is a nodal line along which the intensity drops to zero. This angular dependence can be captured in a polar plot, an example of which is shown in Figure 3.11. In a polar plot, the intensity (or any other value which is a function of angle) is represented as the radial distance from the origin. The line in the plot shows the intensity as a function of angle, and one can see that there is a very strong lobe in the forward direction (the main beam) accompanied by two smaller side lobes (the diffracted beams).

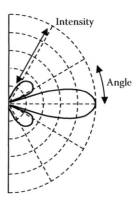

Intensity

Angle

FIGURE 3.11 A representative polar plot. The solid line represents the intensity of the wave as a function of angle.

3.5 Sound Generation and Detection

Sound can be generated by a wide variety of mechanical and electrical devices, including whistles, bells, speakers, piezoelectric crystals, and high-voltage spark gaps. Many of these devices can also be used to detect sound that has been generated elsewhere. The devices that are typically used in process sensor applications are described briefly here.

Electromagnetic speakers, commonly found in radios, consist of a small coil of wire (called a *voice coil*) that is connected to a paper cone or diaphragm. When electrical current passes through the voice coil, it creates a fluctuating magnetic field within the coil. An external magnetic field is generated by a nearby permanent magnet, and the interaction between the two fields produces a force on the coil. This force is transmitted to the diaphragm, which pushes on the surrounding air. Most speakers can be used as microphones, since sound waves can push on the diaphragm and move the voice coil within the external magnetic field. If the field is not uniform, then the changing amount of magnetic flux through the voice coil causes a small voltage to appear across the coil.

Piezoelectric materials such as quartz or barium titanate are ones that generate an electrical charge when placed under a mechanical stress. When an acoustic wave pushes on the surface of the crystal, directly or via a diaphragm that amplifies the applied stress, the piezoelectric element generates a small voltage that fluctuates in response to the acoustic pressure. This effect can also be run backward: the application of a voltage across a piezoelectric element causes the element to constrict by a proportional amount. Sound can be generated by applying a fluctuating signal such as a sine wave to the device, which is usually called a transducer. Therefore piezoelectric transducers can be used as either microphones or speakers. By suitable construction of the transducer, the frequency range of the device can be tailored to a specific application. Piezoelectric transducers are used in audio applications (earphones and small microphones) at relatively low frequencies, but they can also be designed to operate at hundreds of megahertz.

An electret speaker contains a thin membrane that carries a permanent net electrical charge. The membrane is placed between two conductive screens across which an applied voltage creates an electric field. The membrane moves as a result of the resulting force on the captive charges, and sound is produced. Due to the very low mass of the membrane, these speakers have excellent high frequency response. Microphones based on electret membranes are common in computer applications.

3.6 Appendix on the Wave Equation

Airborne sound is composed of pressure waves where the change in local pressure P_Δ measured with respect to equilibrium P_0 is roughly 1 part in 10^7. The change in local pressure causes a small change ρ_Δ in the local density of the air, measured with respect to the equilibrium density ρ_0. These changes are related by the compressibility κ of the air:

$$P_\Delta = \kappa(\rho_\Delta) \tag{3.28}$$

where

$$\kappa = \left(\frac{dP}{d\rho}\right)_0 \tag{3.29}$$

The differential shown above is evaluated at equilibrium. The reason for the change in pressure is that the acoustic wave displaces some of the air molecules, pushing them together near the wavefront and pulling them apart in regions of rarefaction. After the wavefront passes, these molecules are close to their original positions.

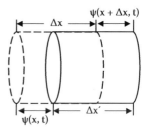

FIGURE 3.12 The volume element of air used in the derivation of the wave equation for sound. The dimension along the x axis is greatly exaggerated for clarity.

Consider a hypothetical volume element of air of thickness Δx and unit area, oriented at right angles to the x axis as shown in Figure 3.12. When a continuous plane acoustic wave traveling along the x axis hits this volume element, it displaces the molecules a distance $\psi(x,t)$, which is a function of position and time (Feynman et al., 2006, vol. 1, pp. 47–4ff). In this new position the element has a density of $\rho = (\rho_\Delta + \rho_0)$ and a new thickness $\Delta x'$ equal to

$$\Delta x' = \Delta x + \Delta x \frac{\partial \psi}{\partial x} \tag{3.30}$$

The mass of a volume element is given by the product of the density and the volume; in this case one side of the volume element has unit area, so the mass equals the product of density and the thickness (this is called the *areal mass*). Assuming that the same number of molecules is in the volume element before and after the deformation caused by the acoustic wave, then it follows that

$$\rho_0(\Delta x) = (\rho_\Delta + \rho_0)\left(\Delta x + \Delta x \frac{\partial \psi}{\partial x}\right)$$

$$= (\rho_\Delta \cdot \Delta x) + \rho_0 \Delta x + (\rho_\Delta \cdot \Delta x)\frac{\partial \psi}{\partial x} + \rho_0(\Delta x)\frac{\partial \psi}{\partial x} \tag{3.31}$$

$$\therefore \rho_0 = \rho_\Delta + \rho_0 + (\rho_\Delta)\frac{\partial \psi}{\partial x} + \rho_0 \frac{\partial \psi}{\partial x} \tag{3.32}$$

Since ρ_Δ and $(\partial \psi/\partial x)$ are small, to first order we can neglect the product of these terms. Furthermore ρ_Δ is negligible compared to ρ_0, so $(\rho_\Delta + \rho_0) \approx \rho_0$; therefore

$$\rho_\Delta = -(\rho_\Delta + \rho_0)\frac{\partial \psi}{\partial x} \approx -\rho_0 \frac{\partial \psi}{\partial x} \tag{3.33}$$

The pressure differential on this volume element is

$$\Delta P = -\frac{\partial P}{\partial x}(\Delta x) = -\frac{\partial P_\Delta}{\partial x}(\Delta x) \tag{3.34}$$

where it is recognized that by definition the P_0 part of P is constant. This pressure differential is the net force per unit area acting on the volume element of air. From Newton's Second Law of Motion we know that an external force acting on an object produces acceleration, and the product of this acceleration and the object's mass equals the applied force. The acceleration of the volume element is $\partial^2\psi/\partial t^2$ and its mass is $\rho_0\Delta x$, so it follows from equation 3.34 that

$$\rho_0(\Delta x)\frac{\partial^2\psi}{\partial t^2} = -\frac{\partial P_\Delta}{\partial x}(\Delta x) \tag{3.35}$$

Dividing both sides of equation 3.35 by Δx and substituting equations 3.28 and 3.33 into it, we find that

$$\rho_0\frac{\partial^2\psi}{\partial t^2} = -\frac{\partial P_\Delta}{\partial x} = -\kappa\frac{\partial\rho_\Delta}{\partial x} = -\kappa\frac{\partial}{\partial x}\left(-\rho_0\frac{\partial\psi}{\partial x}\right) = \rho_0\kappa\frac{\partial^2\psi}{\partial x^2} \tag{3.36}$$

By defining a new constant c (which can be shown to be the wave velocity) through the relation $c^2 = \kappa$, we can rewrite equation 3.36 as

$$\frac{\partial^2\psi}{\partial x^2} = \frac{1}{c^2}\left(\frac{\partial^2\psi}{\partial t^2}\right) \tag{3.37}$$

Equation 3.37 is known as the *wave equation*, and it describes not only the passage of sound through air but also the propagation of waves in general.

Suggested Reading

Halliday, D., Resnick, R., and Walker, J. (2005). *Fundamentals of Physics*, Seventh Edition (chaps. 16-17). New York: Wiley.

Krautkramer, J. and Krautkramer, H. (1983). *Ultrasonic Testing of Materials*, Third Edition New York: Springer-Verlag.

Povey, M.J.W. (1997). *Ultrasonic Techniques for Fluids Characterization*. San Diego, CA: Academic Press.

Strutt, J.W. (Lord Rayleigh). (1945). *The Theory of Sound*, First American Edition. New York: Dover Publications.

4

Light

4.1 Electromagnetic Waves

Scientific inquiry took millennia to ascertain the true nature of light, and it is not possible here to attempt to recapitulate all of the philosophizing, hypotheses, arguments, and experimentation involved in that great debate. The main point of controversy was whether light should be considered as a particle or as a wave. Empedocles of Acragas (492–432 B.C.E.) was of the opinion that light is a "streaming substance" emitted by luminous bodies, and that light travels at a finite speed (Sambursky, 1958). In the seventeenth century, Isaac Newton postulated that light is composed of particles and that color is connected with the size of those particles; however, his contemporary Christiaan Huygens argued that light is a wave that transfers energy but not substance (Einstein & Infeld, 1966, p. 105ff). It turns out that light is indeed a wave propagating via electric and magnetic fields, but in addition to exhibiting properties of waves (such as interference) it also exhibits properties of particles (such as momentum and localization). When treating light as particles, the individual quanta are called *photons*.

Electromagnetic phenomena, including light, are described by Maxwell's four equations, which are easily found in most introductory physics textbooks (see for example Halliday et al., 2005, p. 868). These equations describe the mathematical connections between electrical charge, current, electric field, and magnetic field. The details of these equations and their interpretation are not of immediate concern to us here. However, a particularly interesting result is that these four equations can be combined to produce a relatively simple equation (Feynman et al., 2006, vol. 2, Lecture 18):

$$\frac{\partial^2 \phi}{\partial x^2} + \frac{\partial^2 \phi}{\partial y^2} + \frac{\partial^2 \phi}{\partial z^2} - \frac{1}{c^2}\frac{\partial^2 \phi}{\partial t^2} = -\frac{\rho}{\varepsilon_0} \tag{4.1}$$

where ϕ is the electric potential, ρ is the electrical charge density, and ε_0 is a constant (the permittivity of free space). Note that in the absence of electrical charge, the right-hand side of equation 4.1 becomes zero; in one dimension, we can write the result for a perfect vacuum as

$$\frac{\partial^2 \phi}{\partial x^2} = \frac{1}{c^2} \frac{\partial^2 \phi}{\partial t^2} \qquad (4.2)$$

This result describes the propagation of disturbances in the electromagnetic field through empty space, and it is identical to equation 3.37 (the wave equation) in the previous chapter. Therefore, the solutions to equation 4.2 are the same as the solutions to equation 3.37. Like sound, electromagnetic waves can be characterized by their speed, wavelength, phase, and intensity. The human eye is receptive to these waves provided the wavelength is in the range of approximately 400 to 700 nanometers (nm), and this is what we call *visible light*. The color we perceive is determined by the wavelength of the light.

We have seen that the constant c in the wave equation (equation 4.2) refers to the propagation speed of the wave, and in the case of electromagnetic waves c refers to the speed of light, which is about 3×10^8 m/s in a vacuum.[1] The speed of light is lower in materials than it is in free space by a factor of n, which is the index of refraction of the material. The speed of light c_m in a material is thus

$$c_m = \frac{c}{n} \qquad (4.3)$$

Since the wave speed is slower in materials than in vacuum, the wavelength of light of a given frequency is shorter than it is in free space:

$$\lambda_m = \frac{\lambda}{n} \qquad (4.4)$$

As soon as the light emerges from the material, it resumes its original speed and wavelength.

The wavelength of light is inversely proportional to its frequency, just as in the case of sound:

$$\lambda = \frac{c}{f} \qquad (4.5)$$

Red light, with a wavelength of roughly 600 nm, has a frequency of about 500 THz. X-rays have much shorter wavelengths, and therefore much higher frequency. By contrast, a commercial AM (amplitude modulation) radio station broadcasting at 600 kHz emits radio waves that are 500 meters long.

Light can be classified by its spectral properties. Monochromatic light is light of a single color (only one frequency or wavelength). Such light is produced by electronic transitions within atoms; these transitions (which will be described in more detail in chapter 6) release an energy ΔE [eV] in the form of a photon of light. The wavelength of the photon λ is given by

$$\lambda = \frac{hc}{\Delta E} \tag{4.6}$$

where h is Planck's constant (4.136×10^{-15} eV·s). Obviously the wavelength of the emitted light decreases as ΔE increases. Polychromatic light is a combination of colors; it can be produced when multiple energy levels are involved in the electronic transitions. A hot object emits *blackbody radiation* (discussed in section 7.2.3) that contains a continuous spectrum of wavelengths. White light is an example of a continuous spectrum, although the perception of white light can be produced by combining red, green, and blue light. Monochromatic light is said to be *coherent* when essentially all of the electromagnetic waves are in phase with each other; coherence is the hallmark of laser light. Light that lacks this coordination of phase is described as *incoherent*.

Many sensing applications require a collimated beam of light, which means that the beam maintains a constant diameter over a given distance. Collimation is usually accomplished using lenses or mirrors. The optical cavity in a gas laser is usually long enough to impart some degree of collimation to the beam, but even lasers must be collimated in order to maintain a constant beam diameter.

Another important characteristic of light is its polarization. The oscillating electric and magnetic fields that comprise an electromagnetic wave are always perpendicular to each other and to the direction of propagation of the wave. If the plane defined by the oscillation of the electric field vector and the direction of propagation maintains a constant orientation, the light is linearly polarized; see Figure 4.1a (Halliday et al., 2005, p. 902). If on the other hand the electric field vector rotates clockwise or counterclockwise while maintaining a constant amplitude of oscillation, the light is circularly polarized (Figure 4.1b and Figure 4.1c). Circular polarization is a special case of elliptical polarization, in which the amplitude of the electric field oscillation describes an ellipse (as depicted in Figure 4.1d).

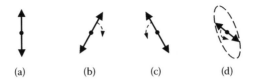

FIGURE 4.1 Polarization of light, where the electric field vector points along the direction indicated if the light ray is viewed head on: (a) linearly polarized light; (b) right-handed circularly polarized light; (c) left-handed circularly polarized light; (d) left-handed elliptically polarized light.

4.2 Optical Elements

This section provides a brief description of many of the optical elements used broadly in scientific and industrial applications. Most of these items are incorporated in the instruments described elsewhere in this book. It should be noted that many of these items have analogs that work with acoustic or ultrasonic waves.

4.2.1 Mirrors

A *mirror* (see Figure 4.2a) reflects light, whether it is a single beam (as from a laser) or an image. In either case, the angle of reflection is the same as the angle of incidence, as mentioned in connection with Figure 4.6a. The mirror works because light is reflected at an interface between materials with different indices of refraction. Typically, mirrors are made of glass, but they may also be made of transparent polymers. In a simple "looking glass" the light is reflected at the back surface, which has been coated with aluminum and painted; in more demanding applications the reflection takes place at the front surface of the mirror, which has been sputtered with aluminum or gold. These front surface mirrors provide a better quality image because the light does not travel through the glass.

Dichroic mirrors are specially designed interference filters (see section 4.2.6) that block long wavelength light but allow shorter wavelengths to pass. The filter is created

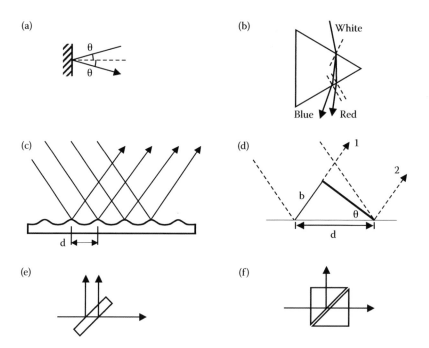

FIGURE 4.2 Optical components: (a) mirror; (b) prism; (c) diffraction grating; (d) the geometry of a diffraction grating; (e) beam splitter; (f) cube beam splitter formed by joining two prisms.

from alternating layers of a transparent coating on a substrate. These layers form a filter that can be tuned to a particular range of wavelengths. Other wavelengths are simply reflected.

4.2.2 Prisms and Gratings

A *prism* (Figure 4.2b) is a wedge of glass or other transparent material. As Newton demonstrated in the seventeenth century, a prism can separate white light into its component colors. The explanation is that the speed of light in the prism depends on wavelength (see the discussion of dispersion at the end of section 3.3). When a beam of light strikes the prism at an angle, it is refracted according to Snell's Law (equation 3.16), in which the index of refraction varies with wavelength. The result is that the light rays are bent; blue light has the higher index of refraction and is therefore refracted through a larger angle than red light. Reflective prisms (not shown in the figure) are also used in optical imaging systems to invert images or to reflect light.

A *grating* (shown in Figure 4.2c) is a transparent substrate (often plastic) that has been etched, embossed, or metallized with a series of fine parallel lines with a spacing between them equal to d. When light strikes a grating, diffraction (see section 3.4.6) causes the light ray to change direction. Adjacent rays are diffracted in the same direction (at angle θ), and they interfere with each other at a large distance from the grating. Figure 4.2d shows two adjacent rays (labeled 1 and 2); ray 1 leaves the grating and travels a distance b before ray 2 emerges. From the geometry of the figure, it is evident that

$$b = d \sin \theta \qquad (4.7)$$

In order for the interference of the two rays to be complete constructive (yielding the maximum light intensity), this distance b must be an integral number of wavelengths λ of the light. Thus we find that the grating diffracts light into certain angles, given by

$$\sin \theta = \frac{m\lambda}{d} \qquad (4.8)$$

where m is some integer $(0, \pm 1, \pm 2, \ldots)$.

Since the diffraction angle is a function of wavelength, such a grating is wavelength dispersive, which means that the light is separated into its component wavelengths. Diffraction gratings therefore serve the same function as prisms but they are flat and can be produced more cheaply.

4.2.3 Beam Splitters

A *beam splitter* is similar to a partially silvered mirror: it splits a beam of light into two by allowing some of the light through and reflecting the rest. The simplest beam splitter is a pane of glass (Figure 4.2e) that may have a coating on it to enhance performance. Beam splitters are often used to create a reference beam that can be compared with scattered or transmitted light returning from a material sample.

A problem sometimes encountered with beam splitters is that the reflection comes from both the front and back surfaces, and the thickness of the glass introduces a separation between the two reflections. Therefore, a single laser beam passing through the beam splitter of Figure 4.2e generates three beams: the transmitted beam and two parallel reflected beams. The extra reflected beam often complicates the alignment and operation of the optical system. To avoid this problem one can use a pellicle, which is a supported membrane only a few micrometers thick. A pellicle operates as a beam splitter, but the front and back surface reflections are essentially collinear due to the thinness of the membrane.

A beam splitter can also be formed by the juxtaposition of two right-angle prisms, which are bonded together with optical cement to form a cube (as depicted in Figure 4.2f). The thickness and refractive index of the cement layer can be designed to adjust the ratio of the split beams or to split a beam into two orthogonally polarized components. Cube beam splitters, like pellicles, do not create a double beam.

4.2.4 Lenses

A *lens* is a device that focuses light (see Figure 4.3a). It operates by refraction, and its shape is designed so that a collimated beam of light is brought to a focus at a distance F (the focal point) from the lens. The focal length can be positive or negative. Lenses have the interesting property that they generate the two-dimensional Fourier transformation of an image, and they are the basis of most optical computing schemes (Goodman, 2005).

Lenses are most often used in imaging applications, where the image of an object is formed. If the distance from the lens to the object is P and the focal length is F, then the image will be formed at a distance Q given by

$$\frac{1}{Q} = \frac{1}{F} - \frac{1}{P}$$

(4.9)

(a)

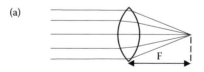

(b)

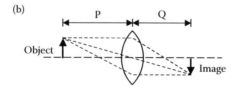

FIGURE 4.3 A simple lens: (a) the focal point F; (b) the formation of an image, where the object is at a distance P and the inverted image is at a distance Q.

Note that the image is inverted; real imaging systems use several lenses to produce an upright image.

4.2.5 Apertures

An *aperture* is simply a hole through which light can pass. In imaging systems, the aperture (usually specified in terms of *f-stop* numbers) limits the cone angle of the light rays that are focused by the lens and so determines the depth of focus. An *iris* is a mechanical shutter (see Figure 4.4a) that forms an aperture, the size of which can be adjusted.

A *pinhole* is an aperture made from a thin metal foil with a precision hole drilled at the center; the diameter of the hole is typically in the range of 1 μm to 100 μm. An important application of pinholes is to provide spatial filtering of a laser beam. Most light sources, including lasers, do not provide a beam with a smooth intensity contour. Optical defects in the system and dust in the air create spatially dependent variations in the light intensity, and these imperfections degrade the quality of the image or measurement. This spatial noise can be filtered out by passing the light through a lens and placing a pinhole at its focus. Since the lens creates a two-dimensional Fourier transformation of the incident light beam, all of the low spatial frequencies are concentrated at the center of the focus. The pinhole removes the higher spatial frequencies, which contain all of the noise components. The resulting light, which can be recollimated with another lens, is more uniform in intensity than the original beam.

4.2.6 Filters

A *filter* is a passive component that modifies light by absorbing or blocking some of its component rays. Perhaps the most familiar type is the color filter, which passes light only at those wavelengths that are in selected regions of the spectrum. In optics, color filters

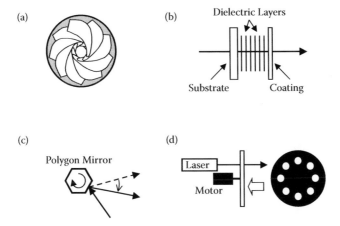

FIGURE 4.4 Optical components: (a) iris; (b) interference filter; (c) rotating polygon mirror; (d) chopper wheel.

are often made from a glass to which pigment has been added. The pigment absorbs light at some wavelengths and reflects light at other wavelengths, with the result that light passing through the pigmented glass takes on a color. A solar filter is one example; it is used in conjunction with charge-coupled device (CCD) cameras to block natural infrared light that would otherwise reduce the contrast in the image.

Interference filters also pass only certain wavelengths of light. Instead of using pigments, they rely on a complex structure of layered dielectric films to produce interference (Figure 4.4b). The refractive index and thickness of the layers are designed to create a resonant structure that is tuned to a range of wavelengths. Light with a wavelength outside the passband is not transmitted very effectively. Interference filters tend to be more expensive than glass filters, but they can be designed to have a very narrow passband (10 nm or less) while retaining a high efficiency (85%).

A neutral density filter is one that reduces the intensity of light equally at all wavelengths over a large portion of the spectrum. The filters are specified in terms of their optical density, which is the negative common logarithm of the transmission coefficient. Therefore, the intensity I_f of the transmitted light as a fraction of the incident light I_0 is given by

$$I_f = I_0 10^{-(OD)} \tag{4.10}$$

Polarizing filters such as Polaroid film allow light to pass only if its polarization is aligned in a particular direction. Although improvements have been made since its invention, Polaroid was originally produced by loading a viscous solution of nitrocellulose with microscopic crystals of herapathite (iodoquinine sulfate) and extruding it so that the needlelike crystals become aligned with each other (Land, 1951; Land & Friedman, 1933). Herapathite crystals have the property that they preferentially absorb light whose polarization is not aligned with the crystal lattice, so the film transmits only those light rays that are linearly polarized in the preferred direction.

4.2.7 Modulators

A modulator is a device that modifies the position, intensity, frequency, or phase of light. Laser scanners, such as bar code scanners used in retail checkout lines, often use a galvanometer mirror (usually called a *galvo mirror*) to sweep the laser beam across the field of view; the galvo mirror is a small mirror attached to an armature or other support that can be rotated by passing electrical current through a small field coil, which generates a magnetic field. The response frequency of the galvo mirror is limited by the mass of the mirror and armature, so a rotating polygon mirror (shown in Figure 4.4c) is often used to achieve very high scan rates.

A simple way to modulate the intensity of a light beam is to pass it through a *chopper wheel*, which is a disc with a series of holes around its perimeter. The disc, which is turned with a small motor, is painted flat black to minimize reflection. As the disc rotates, it periodically blocks the beam (as depicted in Figure 4.4d). This modulation is necessary to implement phase-sensitive detection schemes, wherein the detector signal from a given measurement is analyzed by passing it through a phase-locked loop tuned to the frequency of the modulation. The component of the signal that is exactly synchronized with the light modulation is the desired measurement, whereas the rest of the detector signal is due to noise. Instruments that perform this analytical function are called lock-in amplifiers; they are used to make very precise measurements.

The frequency of a light beam can be modulated by using a small mirror mounted on a piezoelectric element (or some other mechanism) that can push the mirror back and forth (Figure 4.5a). Since the surface of the mirror is moving relative to the light source, the reflected light experiences a Doppler shift in its frequency; equation 3.25 gives the change in frequency of the light at the point where it impinges on a receding mirror. When the mirror is approaching the source, the light is blue shifted so the frequency increases. Since the mirror is also moving relative to the stationary detector, the frequency shift is doubled. Therefore, the frequency of the reflected light f_r is given by

$$f_r = \left(\frac{1+\beta}{1-\beta} \right) f_0 \approx (1+2\beta) f_0 \qquad (4.11)$$

where f_0 is the frequency of the incident light, $\beta = v/c$, c is the speed of light, and v is the velocity of the mirror toward the source and detector. The change in optical frequency Δf is given by

$$\Delta f = f_0 - f_r = 2\beta f_0 = \frac{2v}{c} f_0 \qquad (4.12)$$

If the light beam is not at normal incidence (i.e., at right angles to the mirror), then the component of the velocity parallel to the beam is used. The drawback of this approach is that the device pushing the mirror has a limited range of motion, so a constant shift in frequency cannot be maintained indefinitely.

A device that can provide a sustained shift in frequency is the *acousto-optic modulator*, which is also called a *Bragg cell*. Figure 4.5b depicts the operation of a Bragg cell, in which an ultrasonic transducer sets up traveling waves in a transparent material such as quartz. These waves create density variations at regular intervals that equal the ultrasonic wavelength. Due to the change in refractive index, each of these successive waves acts as a partially reflective mirror moving with velocity v. The light is reflected by each

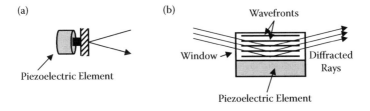

FIGURE 4.5 Modulators: (a) piezoelectric element pushing on a mirror; (b) Bragg cell.

of the wavefronts at the sample angle θ as the angle of incidence. These beams must all be in phase with each other or else destructive interference will cause the beam to disappear. This requirement leads to the diffraction condition

$$2\lambda_s \sin\theta = \frac{\lambda}{n} \tag{4.13}$$

where λ_s is the wavelength of the sound in the Bragg cell, λ is the wavelength of the light in vacuum, and n is the refractive index of the cell material (Yariv, 1985, p. 387ff).

The ultrasonic wave in a Bragg cell moves with velocity v, so the component of velocity in the direction of the light beam is $v \cdot \sin\theta$. From equation 4.12 we see that the change in optical frequency is therefore

$$\Delta f = \frac{2v\sin\theta}{c}f_0 \tag{4.14}$$

An expression for $\sin\theta$ can be obtained from equation 4.13, and after substitution into equation 4.14 we find that

$$\Delta f = \left(\frac{2v}{c}\right)\left(\frac{\lambda}{2\lambda_s n}\right)f_0 = \left(\frac{v}{\lambda_s}\right)\left(\frac{\lambda f_0}{cn}\right) = \left(\frac{v}{\lambda_s}\right) = f_s \tag{4.15}$$

where f_s is the frequency of the ultrasonic wave. Equation 4.15 proves that the optical frequency of the diffracted light beam is shifted by the frequency of the ultrasonic wave passing through the Bragg cell. Although a given wave front in the traveling ultrasonic wave eventually reaches the end of the cell, the wave itself continues to emanate from the transducer, and the frequency shift can be maintained indefinitely. The ability to change the optical frequency of light enables various detection schemes that use optical heterodynes.

4.3 Light Generation and Detection

4.3.1 Incoherent Sources

Most light sources are *incoherent*, which means that the light waves are randomly produced by the source and do not have any fixed phase relationship between them. Natural sources of light produce a spectrum of light known as *blackbody radiation* (see section 7.2.3), which is emitted by matter at high temperature. Stars produce light as a by-product of thermonuclear reactions; wood fires, candles, and gas lamps produce light as a by-product of the oxidation of fuel. Among artificial sources, the most familiar light is Thomas Edison's electric incandescent lamp, which generates light (plus a good deal of heat) by passing current through a filament. The resistive losses in the filament cause it to become white hot, and the resulting blackbody radiation covers a broad portion of the electromagnetic spectrum, including visible light (Figure 7.2).

Gas-discharge lamps, such as those found in neon signs, use a high voltage source to ionize gas within the lamp; the ions are accelerated by the electric field and cause collisional excitation and additional ionization of the atoms in the gas. As the atoms decay back to their ground state, they emit photons whose wavelength is determined by difference in energy level ΔE between the two electronic states of the atom according to equation 4.6. The emitted light is therefore composed of distinct wavelengths that are characteristic of the atoms in the gas; the spectral output of such a lamp is therefore very different from that produced by an incandescent source. Fluorescent lights are gas-discharge lamps with a coating on the inside surface of the glass tube; this coating absorbs ultraviolet light and fluoresces in the visible spectrum.

A *strobe light* is a special type of gas-discharge lamp in which the applied voltage is insufficient to ionize the gas in the tube. A trigger electrode mounted near one end of the tube is used to create a group of ions that are immediately accelerated by the electric field in the tube. As these ions strike other gas molecules in the tube, those molecules are ionized, and within a few microseconds the gas in the entire lamp has been ionized. The ions quickly recombine with free electrons, emitting photons in the process. The result of this process is that a very short and very bright pulse of light is generated. Strobe lights are used to illuminate objects that are moving rapidly. The short duration of the illumination makes it possible to capture a clear image of the object regardless of a camera's shutter speed.

Solid-state sources, such as *light-emitting diodes* (LEDs), rely on a somewhat different principle of operation. As discussed in section 5.2.1, an LED produces light when the charge carriers (electrons and holes) flowing through a semiconductor junction recombine with each other. This recombination represents a drop in energy level for the electron, so the excess energy is removed through the creation of a photon of the corresponding energy given by equation 4.6.

4.3.2 Lasers

A *laser* is a coherent, monochromatic light source that operates by stimulated emission, which is an induced transition, between a higher energy level E_2 and a lower one, E_1. Induced transitions between two electronic states in an atom or molecule are caused by the passage of a photon with an energy equal to the difference

$$\Delta E = E_2 - E_1 \tag{4.16}$$

If the atom is in the lower state (1), then a finite probability exists that the atom will absorb the photon and switch to the upper state (2). However, if the atom is already in state 2, then it can be stimulated by the photon to drop to state 1, releasing a photon in the process (Yariv, 1985, p. 130ff). Both of these photons can continue to stimulate further emission of light, and due to the synchronization between passage of the light wave and additional emission, the light produced is coherent (i.e., in phase).

If a collection of atoms is at thermal equilibrium, then the number of atoms in the upper and lower states (denoted by N_2 and N_1, respectively) are related by

$$\frac{N_2}{N_1} = e^{-\Delta E / kT} \tag{4.17}$$

where k is Boltzmann's constant (8.617×10^{-5} eV/°K) and T is the absolute temperature [°K]. Because equation 4.17 shows that $N_2 < N_1$ at thermal equilibrium, it follows that a beam of light passing through a gas (for instance) will be absorbed rather than enhanced. If, on the other hand, the population of the states is inverted so that $N_2 > N_1$, stimulated emission can occur and the beam will be amplified.

The trick to creating a laser is to produce an inversion of the normal population of electron states; this inversion may be produced by exciting the atoms or molecules to levels that are metastable—that is, that do not readily decay to the ground state. The metastable state (corresponding to level 2) can be populated by pumping electrical or optical energy into the system. When a population inversion is produced in a transparent medium that is contained within an optical resonator (as shown in Figure 4.6), the light from spontaneous transitions to the lower state will stimulate additional transitions to the lower state. The resulting light in the optical resonator will tend to be collimated (since the rays bouncing back and forth between the mirrors will be amplified more than the others), and it will be coherent. The length of the resonant cavity is tuned to match the wavelength of the light, and one of the mirrors in the cavity is designed to transmit 1% or so of the light, which emerges as a laser beam. Additional details are discussed in Yariv (1985).

Similarly, a population inversion can be created in a semiconductor by injecting a large number of electrons into the conduction band. As these electrons fall back into the valence band, they stimulate the coherent transition of other conduction electrons. Since semiconductor lasers are very small, their output is quite divergent, and specially designed lenses must be used to collimate the beam.

4.3.3 Detectors

Light can be detected using a variety of sensors. The most common ones are photoconductive detectors, photodiodes and phototransistors, and photon multiplier tubes (PMTs). Photoconductive detectors such as photoresistors are semiconductor crystals across which an electric field is applied. Since semiconductors do not contain many electrons in the conduction band (see chapter 5), they will ordinarily not conduct current. However, if a photon of sufficient energy is absorbed, a valence band electron can be promoted to the conduction band, where (in the case of the n-type semiconductor) it is available to ferry charge across the device. A "hole" or electron vacancy is left behind in the valence band when the electron is promoted. Eventually the electron recombines with a hole and falls back into the valence band, so a steady source of light is needed to

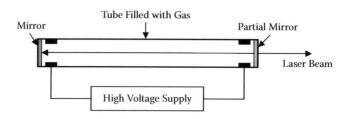

FIGURE 4.6 Sketch of a gas-filled laser.

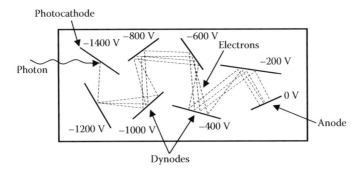

FIGURE 4.7 Structure of a photomultiplier tube.

produce a steady electrical current through the detector. Since the number of electron-hole pairs created depends on the number of photons, it is evident that the current [A] is proportional to the power [W] of the beam.

A photomultiplier tube, shown in Figure 4.7, is an extremely sensitive device consisting of a photocathode, a series of electrodes (dynodes), and an anode enclosed in a glass vacuum tube. A set of electric fields is established by connecting the photocathode, dynodes, and anode to a power supply network so that the voltage between adjacent dynodes is 100–200 volts (V). The photocathode is made of metal alloys with very low work functions that are less than about 2 eV. The work function E_W is the minimum amount of energy necessary to pull a conduction band electron off the surface of a given metal. If a photon of wavelength λ strikes the metal, then the maximum kinetic energy of the ejected electron (called the photoelectron) E_K is

$$E_K = \frac{hc}{\lambda} - E_W \tag{4.18}$$

Equation 4.18 describes the photoelectric effect (Einstein, 1905). When $E_K > 0$, some of the photons striking the photocathode will cause photoelectrons to be ejected; once in the electric field, the photoelectrons are accelerated toward the nearest dynode.

Upon striking the dynode, the accumulated kinetic energy of the incident photoelectron is equal to the potential energy difference between that dynode and the preceding one. Since this energy is many times greater than E_W, a shower of secondary electrons is created. These are accelerated to the next dynode where each one in turn generates another burst of electrons. If the dynode structure includes enough stages, the electronic amplification factor can exceed 10^7. The final shower of electrons eventually reaches the anode and appears as current flowing to the anode power supply.

The PMT can be operated in two separate modes. In *current mode*, the output current appearing at the anode is proportional to the intensity of the incident light. In *counting mode*, the dynode voltages are increased to maximize sensitivity, and the individual bursts of current are counted. Counting mode is used to measure extremely low levels of light.

The operation of phototransistors and CCD imaging detectors is covered in the following chapter.

Suggested Reading

Haber-Schaim, U., Dodge, J.H., Gardner, R., Shore, E.A., and Walter, F. (1991). *PSSC Physics*, Seventh Edition (chaps. 17-19). Dubuque, IA: Kendall/Hunt.

Yariv, A. (1985). *Optical Electronics*, Third Edition (chaps. 4-6, 11, and 15). New York: Holt Rinehart and Winston.

5

Electricity and Electronic Devices

The previous chapters have introduced sound, light, and ionizing radiation as physical phenomena that interact with their environment; these interactions provide the physical observables needed to detect changes in the process or product. Electrical phenomena also respond to changes in the surrounding environment and provide additional mechanisms that can be used in process sensors. Electrical measurements are in fact crucial to process sensors because ultimately all sensors rely on electricity and electronic readouts. For this reason, key concepts in electrical measurement and electronic circuits are reviewed in this chapter.

5.1 Electricity

5.1.1 Potential and Current

Electrical phenomena (e.g., static electricity, conduction, and resistance) are understood in terms of charge, potential, and electric field. Although it is difficult to talk about one of these fundamental concepts without invoking the other two, they are distinctly different from each other. The concept of electrical charge is defined in terms of the force on a particle with a charge q_1 when placed in proximity to another charge q_0. Coulomb's Law states that the force F_C is

$$F_C = \left(\frac{1}{4\pi\varepsilon_0} \right) \frac{q_1 q_0}{r^2} \tag{5.1}$$

where ε_0 is a constant (called the permittivity of free space) and r is the distance between the two charges. Equation 5.1 bears an obvious similarity to Newton's Law of Gravitation, which states that the force F_G between two masses m_1 and m_0 is given by

$$F_G = G \frac{m_1 m_0}{r^2} \tag{5.2}$$

49

where G is the gravitational constant. Thus charge might be regarded as an electrical analog of mass. Unlike mass, however, charge has two polarities: positive and negative. In equation 5.1, charges of opposite sign attract each other, and charges of the same sign repel each other. Readers will recall that electrons carry a negative charge and protons carry a positive charge. There are also negative and positive ions, and most particulate matter carries a mixture of positive and negative charges on the surface. The unit of measurement of charge is the coulomb (C), and an electron has a charge of 1.602×10^{-19} C; this quantity is called the *elementary charge* because in classical physics charge is quantized and only appears as an integral multiple of the elementary charge.

When a mass m is placed in the gravitational field of the earth, it is accelerated downward (toward the Earth) at a rate of 9.8 m·s^{-2}. In the same manner, a single positive electrical charge q placed in an electric field E experiences a force F_E equal to

$$\mathbf{F}_E = q\mathbf{E} \tag{5.3}$$

where the direction of the force is in the direction of the applied electric field (hence the use of boldface letters, which denote vectors). In general, the electric field can be a function of position and time. By comparing equations 5.1 and 5.3, it follows that a point charge q_0 creates a position-dependent electric field of strength $E_0(r)$ around itself:

$$E_0(r) = \left(\frac{1}{4\pi\varepsilon_0} \right) \frac{q_0}{r^2} \tag{5.4}$$

Electric field strength is measured in volts per meter (V/m).

To follow the analogy with mass a little further, just as lifting a massive object in a gravitational field requires work, so moving an electrical charge q against the force given in equation 5.3 also requires work. For rectilinear motion in the x direction, work is defined to be the energy required to push against a force F through a given distance X, and this work is equal to the change in the potential energy, ΔU:

$$\Delta U = \int_0^X F(x)dx = q \int_0^X E(x)dx \tag{5.5}$$

Work is measured in joules (J). This work must be supplied by an outside agency (i.e., whatever is causing the charge to move against the force), which essentially stores potential energy in the electric field. It turns out that the integral of $E(x)$ over a distance between two points is equal to the difference in electrical potential ϕ (not to be confused with the phase offset φ used in chapter 3) between them:

$$\Delta U = q(\Delta\phi) \tag{5.6}$$

Electric potential is measured in volts (V); the zero point of the voltage scale is usually taken to be the potential of the earth ("ground"). The relation between the field and the potential is given by

$$E(x) = -\frac{d\phi}{dx}$$

(5.7)

The storage of energy implied by equation 5.6 is reversible; when, for example, an electron moves from a high-potential region to a low-potential region, the energy difference appears as an increase in the electron's kinetic energy (or enables it to do work). An electron that drops through a potential difference of 1 V can do an amount of work equal to 1 electron volt (eV), which is 1.602×10^{-19} J.

Like the electric field, the electric potential is in general a function of position. In the general case of three-dimensional motion, equation 5.5 becomes

$$\Delta U = q \int_{path} E(\mathbf{r}) \cdot d\mathbf{r}$$

(5.8)

where the field forms a scalar product with the differential element $d\mathbf{r}$, which is a vector pointing along the path taken by the moving charge. The three-dimensional form of equation 5.7 is

$$\mathbf{E}(\mathbf{r}) = -\nabla \phi$$

(5.9)

It should be noted that the potential is a scalar function of position. A well-known consequence of equation 5.9 is that, regardless of the path taken in the calculation of equation 5.8, if the path closes on itself to form a complete circuit the ΔU and $\Delta \phi$ will both equal zero (Jackson, 1975, p. 35). Another way of stating this result is that if a closed circuit is split into several sections and the potential difference across each section is measured, then the sum of all potential differences will equal zero. This result is *Kirkoff's Second Law*, and it is used in circuit analysis.

Electrons have been mentioned above as charge carriers, and in electrical wires they are effectively the only charge carriers since the atomic nuclei are locked into place in the crystal lattice of the metal. Most of the electrons are trapped in orbit around their respective nuclei; these are called valence band electrons, and their energy levels are quantized. A few of the electrons at the higher energy levels can become dissociated from their respective nuclei due to thermal energy; these electrons belong to the conduction band and are able to drift throughout the metal lattice.[1] In metals, the conduction band lies just above the valence bands, but in semiconductors there is an energy gap between them (Figure 5.1). The application of an electrical potential across a length of wire creates an electric field that causes the conduction band electrons to drift toward the positive end of the wire. The charge conveyed by the electrons is electric current.

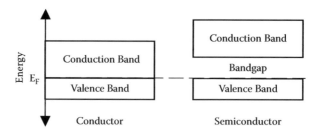

FIGURE 5.1 Band structure in conductors and semiconductors. All of the valence states lie in energy bands just below the Fermi energy, E_F. In semiconductors, there is an energy gap between the valence bands and the conduction band.

The current I is the net flow of charge through a given surface (usually the cross-section of the electrical wire) divided by time:

$$I = \frac{dq}{dt} \tag{5.10}$$

The unit of current is the ampere (A), which is equivalent to one coulomb per second. The total charge that flows through the wire is therefore equal to the current multiplied by the amount of time. Although the charge carriers are electrons (which carry negative charge), the convention is to think of the current as moving in the direction a positive charge would move; therefore conventional current flows from the anode to the cathode.

Steady-state current flow is possible only in a complete circuit. Figure 5.2a shows a schematic of a circuit that consists of a battery and a loop of wire. Current is pushed through the wire by the potential provided by the battery; if the circuit is opened, as by the switch shown in Figure 5.2b, then the flow of current stops. Time-varying electric fields can propagate through space, but direct current (which flows in one direction) requires a conductive path.

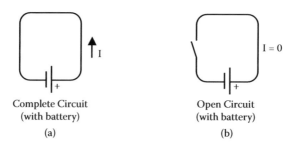

FIGURE 5.2 Simple circuits: (a) a closed (complete) circuit consisting of a loop of wire and a battery in which current I is flowing counterclockwise; (b) an open (incomplete) circuit in which the switch (shown in the off position) prevents current from flowing.

If the current $I(t)$ is fluctuating as a function of time, then the total amount of charge conveyed is given by

$$q = \int I(t)dt \tag{5.11}$$

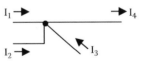

A consequence of equation 5.11 is that the total current flowing into a junction of two or more wires must equal the total current flowing out of the junction, as depicted in Figure 5.3; otherwise, the total electric charge would not be conserved. This observation is a general rule (*Kirkoff's First Law*) that can be used for circuit analysis.

FIGURE 5.3 According to Kirkoff's First Law, the net flow of current into a circuit node must be zero, so the currents shown here must satisfy the equation $I_4 = I_1 + I_2 + I_3$.

Although electricity travels through wires very quickly, the drift speed of the conduction electrons is actually quite slow (on the order of 10^{-7} m/s). To estimate the drift speed, assume that the application of a given electric field causes the electrons drift at a constant speed v_d. The charge in a wire segment of length L and cross-sectional area A is ($neAL$), where the product (ne) is the carrier charge density [$C \cdot m^{-3}$]. This amount of charge drifts through a given cross-section in a time interval equal to (L/v_d). It follows that

$$I = \frac{q}{t} = neAL\left(\frac{v_d}{L}\right) = neAv_d \tag{5.12}$$

$$\therefore v_d = \frac{I}{(ne)A} \tag{5.13}$$

As a concrete example, equation 5.13 predicts that the drift speed in a 1.8 mm diameter copper wire carrying 17 mA of current is about 5×10^{-7} m/s (Halliday et al., 2005, p. 687). The drift speed is therefore very slow, even though the transmission of energy appears to be nearly instantaneous; the explanation of this paradox is that energy is quickly transferred between conduction band electrons as they collide.

Before moving on to electronic devices, it should be mentioned that when current passes through a wire, it creates a magnetic field around the wire. The strength of the magnetic field is proportional to the current. If the wire is wound around a core to form a solenoid, then the strength of the magnetic field in the middle of the solenoid is also proportional to the number of turns in the winding. The voice coils found in speakers (see chapter 3) are solenoids, and when they are activated with alternating current they actuate the speaker to create sound. Beyond this use, magnetism does not play a role in any of the specific sensors described in this book, so a discussion of this topic is omitted here.[2]

5.1.2 Resistance

When an electric potential is applied across a wire, current flows through the wire. If the same potential is applied across a second wire made of a different material than the first, one generally finds that the currents flowing in the two wires are not equal. The resistance R of a wire is defined by the equation

$$R = \frac{V}{I}$$
(5.14)

where V is the potential difference (voltage) and I is the current flowing through the wire or device. The unit of resistance is the ohm, which is equal to 1 V of potential divided by 1 ampere of current. A *resistor* is a two-terminal device with a fixed resistance.

The resistance of a wire of length L and cross-sectional area A is

$$R = \left(\frac{L}{A}\right)\rho$$
(5.15)

where the resistivity ρ is an intrinsic property of each material. The unit of resistivity is the ohm meter ($\Omega\cdot$m). Conductivity σ is also a material property, defined as the reciprocal of resistivity:

$$\sigma = \frac{1}{\rho}$$
(5.16)

The unit of conductivity is $\Omega^{-1}\cdot$m^{-1}, which is also called siemens per meter (S/m); it is also common to find conductivity written in terms of siemens per millimeter (S/mm).

Resistivity and conductance vary with temperature by an amount that depends on the temperature coefficient α:

$$\Delta\rho = \rho_0\alpha(\Delta T)$$
(5.17)

Therefore, a resistor that is made from a material with a known temperature coefficient of resistivity can be used to measure the temperature; such devices are called *thermistors* because their resistance changes in a predictable way with temperature.

Figure 5.4a shows the schematic of a simple circuit consisting of a variable voltage power supply and a resistor. As the voltage is changed, the current flowing through the loop also changes according to

$$I = \left(\frac{1}{R}\right)V$$
(5.18)

The current is plotted in Figure 5.4b as a function of voltage; the graph is a straight line through the origin with slope equal to $1/R$. Ohm's Law, which is usually written as

$$V = IR$$
(5.19)

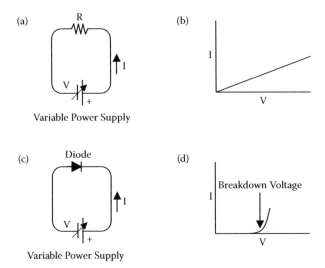

FIGURE 5.4 Ohmic versus nonohmic response: (a) a simple circuit with a resistor and an adjustable voltage supply yields (b) a linear plot of current versus voltage; (c) a reverse-biased diode circuit conducts negligible current until the diode's breakdown voltage is reached, as depicted in (d).

asserts that the current is always proportional to the applied voltage, and devices for which this assertion is true are called *ohmic*. Many devices, such as diodes and transistors, do not obey Ohm's Law. An example is the reverse-biased diode shown in Figure 5.4c; virtually no current flows through the circuit until the applied voltage exceeds the breakdown potential, at which time the current soars (Figure 5.4d). Such devices are described as *nonlinear* or *nonohmic*.

5.1.3 Capacitance and Inductance

A *capacitor* is a device that can store electrical charge; when current flows through a capacitor, charge builds up within the capacitor and creates a voltage drop V_C (and therefore an electric field) across it. The amount of charge Q stored in the capacitor is proportional to the voltage drop:

$$Q = CV_C \tag{5.20}$$

The proportionality constant is the capacitance C, which is a measure of the total charge capacity. Capacitance is measured in farads (F), and one farad is equal to one coulomb divided by one volt. The operation of the capacitor circuit of Figure 5.5a is shown in Figure 5.5b; when the switch is turned on to complete the circuit, charge accumulates in the capacitor, and the value of V_C increases until it finally equals the voltage

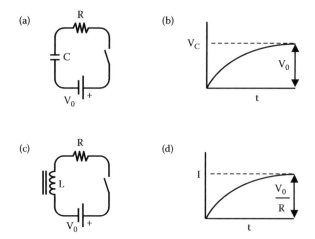

FIGURE 5.5 Charging time in capacitor and inductor circuits: As the capacitor in circuit (a) charges, the potential across it (V_C) asymptotically approaches the applied voltage V_0, as shown in (b). The inductor in circuit (c) initially limits the current I flowing through it, but the current increases with time as shown in (d) and asymptotically approaches the maximum current (V_0/R).

supply V_0. From equation 5.18 it should be apparent that the current through the resistor (and therefore through the entire circuit) is given by

$$I = \frac{(V_0 - V_C)}{R} \tag{5.21}$$

When $V_C = V_0$, the current drops to zero and the capacitor stops charging. Since current is the differential with respect to time of the charge Q, one can differentiate equation 5.20 to find that the current is proportional to the rate of change of the voltage:

$$I = C\frac{dV_C}{dt} \tag{5.22}$$

By combining equations 5.21 and 5.22 and solving for V_C, it can be shown that the voltage as a function of time is

$$V_C = V_0(1 - e^{-t/RC}) \tag{5.23}$$

The product RC is known as the *RC time constant*; if R is the resistance in ohms and C is the capacitance in farads, then RC is the time in seconds until the voltage reaches 63% of V_0. The time constant determines how quickly the capacitor can respond to changes in signal level, and the circuit shown in Figure 5.5a is a type of low-pass frequency filter that removes high frequency content from a signal.

Capacitors store charge in two electrodes (one positive, one negative) that are separated by a thin dielectric material. A *dielectric* is a nonconducting material that is polarizable, which means that bound charges within it can rearrange themselves to minimize the electric field. The dielectric constant is a material property that is closely related to the polarizability. For a simple parallel plate capacitor, the voltage drop across the electrodes (and therefore across the capacitor) is equal to the electric field [V/m] times the thickness of the dielectric layer m. By choosing a material with a higher dielectric constant, the electric field and therefore V_C are both minimized for a given amount of charge. From equation 5.20 it should be evident that the capacitance of such a device is greater than that of the original capacitor. In general, the capacitance for a given electrode geometry is proportional to the dielectric constant.

Electrical capacitance is a general electrical phenomenon that is not limited to capacitors. Capacitance is also exhibited between wires, within circuit boards, and even by the human body (the proof is the familiar shock from buildup of static electricity produced by walking across a rug). This unwanted capacitance is called *stray capacitance* and is on the order of picofarads; it creates alternate, sometimes disruptive, signal pathways in high-frequency circuits. Measurements of the capacitance between two electrodes can also be used to determine the dielectric constant of material placed between them. This type of measurement is used in chapter 9 as the basis for generating tomographic images of the grinding bead distribution inside a media mill.

An *inductor* is essentially a coil of wire that generates a magnetic field when current passes through it (similar to an electromagnetic solenoid). As the magnetic field is established within the coil, a self-induced potential V_L appears across coil that opposes the applied voltage:

$$V_L = -L\frac{dI}{dt} \tag{5.24}$$

The constant L in the equation is called the *inductance*, and its unit of measure is the henry (H). The circuit of Figure 5.5c shows an inductor connected to a resistor, a switch, and a battery. The time dependence of the current in the circuit (after the switch is turned on) is shown in Figure 5.5d.

Capacitors and inductors are both reactive, which means that their effect on time-dependent signals depends on frequency. In the case of a sinusoidal signal, a capacitor behaves as a frequency-dependent resistance

$$R_C = \frac{1}{\omega C} \tag{5.25}$$

but the phase of the current leads the phase of the voltage by ($\pi/2$) radians. Similarly, an inductor behaves as a frequency-dependent resistance

$$R_L = \omega L \tag{5.26}$$

but the phase of the current lags the phase of the voltage by ($\pi/2$) radians. For a true resistor, the current and voltage are always in phase. The combination of resistance and reactance is called *impedance*, and in general it is a function of frequency.

In order to keep track of phase, electronic circuit analysis is often based on the use of complex numbers (called *phasors*) that represent the voltage and current of an alternating-current signal.[3] In this notation, the time dependence is represented by the complex factor ($e^{i\omega t}$), and the impedance Z of the resistor, capacitor, and inductor are written as

$$Z_R = R \tag{5.27}$$

$$Z_C = -\frac{i}{\omega C} \tag{5.28}$$

$$Z_L = i\omega L \tag{5.29}$$

where i is the imaginary number. With these definitions, Ohm's Law can be written as

$$\mathbf{V} = \mathbf{IZ} \tag{5.30}$$

and used to analyze voltage drops and current flow as a function of the angular frequency ω. It is found that electrical signals are partially reflected at points in the circuit where the impedance changes; in this respect the frequency-dependent impedance is analogous to the wavelength-dependent refractive index in optics.

5.2 Semiconductor Devices

5.2.1 Diodes

A *semiconductor* is an otherwise insulating material that has been doped with impurities to create charge carriers that can conduct current. Depending on the type of added impurity, the semiconductor can be either *n-type*, which can donate negative charges (electrons), or *p-type*, which can accept them. When a junction is formed between n-type and p-type semiconductors, the available electrons from the n-type region diffuse into the p-type region, where they lose their mobility.[4] This process is self-limiting, because for every electron that leaves the n-type region, a positive ion is left behind; eventually the growing electric field prevents additional electrons from leaving the n-type region. Without mobile charge carriers, the p-n junction itself becomes an insulator.

If the junction is biased by applying a voltage across it in such a way that the p-type region is at a higher potential than the n-type, then the electric field created by recombination is overcome and the electrons are able to cross the junction. This configuration is called *forward bias*. If the junction is reverse-biased so that the p-type region is at the same or lower potential than the n-type region, then no current can flow. If the reverse bias is sufficiently strong, then breakdown occurs and current is forced through the junction, as depicted previously in Figure 5.4d. The semiconductor device just described is a *diode*, and the schematic symbol for it has been shown in Figure 5.4c.

Light emitting diodes (LEDs) emit light when they conduct current. This emission is due to the energy liberated as the conduction electrons recombine with holes in the p-type region. The wavelength of the light can be customized by adjusting the dopants so that the change ΔE in energy level corresponds to the desired wavelength. The wavelength λ of the emitted light is then equal to

$$\lambda = \frac{hc}{\Delta E} \tag{5.31}$$

where h is Planck's constant (4.136×10^{-15} eV·s) and c is the speed of light (2.998×10^8 m/s). The symbol for the LED is shown in Figure 5.6.

5.2.2 Transistors

A *bipolar transistor* is a semiconductor device with three terminals (base, emitter, and collector), as depicted by the schematic symbols in Figure 5.6. *PNP transistors* are formed by a layer of n-type semiconductor (the base) between two layers of p-type semiconductor (the emitter and collector). *NPN transistors* are the reverse polarity: they are formed by a layer of p-type semiconductor (the base) between two layers of n-type semiconductor (the emitter and collector). These devices are biased so that the potential at the base is midway between the potential at the collector and the potential at the emitter. This arrangement ensures that the base-emitter junction is forward biased, which injects charge carriers into the base region. The collector-base junction is reverse biased, so in the absence of the extra charge carriers no current would flow; however, the injected charges provide conduction between the collector and the emitter. In this manner a small change in current flowing into the base creates a relatively large change in the current flowing through the transistor; the ratio of the collector current to the base current is usually denoted as β. Bipolar transistors are therefore devices that provide a gain in current, with the collector acting as a source (or sink) of current. Transistors are the heart of amplifier circuits, which are discussed in the following section.

The *field effect transistor* (FET) differs from the bipolar transistors described above in that it operates as voltage amplifier rather than a current amplifier. The structure of an FET contains a source, a gate, and a drain; these elements are analogous to the emitter, base, and collector. The source and drain are opposite ends of a semiconductor bar (either n-type or p-type), which is straddled by the gate; depending on the design of the FET, the gate electrode may be insulated from the semiconductor or diffused into it. The current

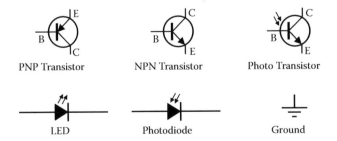

FIGURE 5.6 Common symbols used in drawings (schematics) of electronic circuits.

flowing from source to drain in an FET is largely confined to a channel whose width is determined by the electric field from the applied gate voltage. The gate does not draw any current, and in general FETs operate at very low power levels.

The *phototransistor* (Figure 5.6) relies on light, rather than electrical current, to cause conduction. Its design includes an optical window that allows light to strike the base; the photoelectric effect creates charge carriers in the base region, so the amount of current flowing through the collector is proportional to the amount of light striking the phototransistor. A similar device is the *PIN photodiode*, which is a sandwich of p-type semiconductor, intrinsic semiconductor (with no free charge carriers), and n-type semiconductor. The PIN diode is normally reverse biased, so no current flows through the diode until light reaches the depletion zone that extends throughout the intrinsic layer. These devices are both used as optical detectors.

5.2.3 Integrated Circuits

Semiconductor devices are combined in circuits with other electronic components to perform specific functions. Depending on the complexity of the circuit, the connections may be on a circuit board, on a single piece of silicon (the integrated circuit, or IC), or in hybrid packages that contain several interconnected large-scale ICs.

An example of a large-scale IC is the charge-coupled device (CCD) camera, which captures spatial patterns of illumination. A CCD contains an array of millions of individual light-sensitive detectors that correspond to the *pixels* ("picture elements") in the final image. When light strikes one of the detectors, the photoelectric effect produces electrical charge that is collected and stored locally. The amount of stored charge depends on the intensity of light at that point, so the pattern of illumination is recorded as a pattern of stored charges in the array. The detectors are electrically connected in rows so that the charge can be moved from one to the next; by repeating this charge transfer process a number of times, the entire pattern of charges can be read out of the row. A similar process is used to gather information from the end of each row so that the entire image that can be read out is read out sequentially.

The elements in a CCD sensor register the intensity, not the color, of the light. In order to detect color, a composite color filter (containing the three primary colors) is placed over the sensor. This filter divides the detector elements into triads; each of these elements is sensitive only to red, blue or green light. By combining the intensity data from all three sensors, the full color at that point in the image can be captured. A similar technique is used in color displays, where the close proximity of red, blue and green dots on the screen conveys the sense of color.[5]

5.3 Amplifiers

The electrical signals produced by most sensors tend to be small (e.g., a few millivolts of potential or a few microamperes of current). It is therefore usually necessary to amplify the signal so that the output voltage is (in the case of a voltage input)

$$V_{out} = GV_{in}$$

(5.32)

or (in the case of a current input)

$$V_{out} = HI_{in} \qquad\qquad (5.33)$$

Here G is a voltage gain factor and H is a transimpedance gain factor [V/A]. An ideal amplifier is a circuit that functions according to equation 5.32 or 5.33. Real amplifier circuits often have a slight offset voltage that must be adjusted in order to bring the baseline voltage to zero.

The operation of a basic amplifier is illustrated in Figure 5.7a; the power supply is not shown, but by convention it is assumed to have one terminal grounded (in this case, the negative one). When a signal is applied to the input, it varies the amount of current flowing through the base to the emitter. This variation in base current causes a proportional change in the current collector, where the proportionality constant is the β factor of the transistor (a typical factor is 50). Most of the collector current flows through the 10 kΩ output resistor. The output voltage equals 15 V (the supply voltage) minus the voltage drop across the output resistor; from equation 5.19 it is evident that increasing collector current decreases the voltage at the output of the amplifier. Thus the output signal is stronger than the input signal, but it has the opposite sign; more sophisticated amplifier designs provide a noninverted output signal. The schematic symbol for an amplifier is shown in Figure 5.7b.

An operational amplifier ("op-amp") is a versatile circuit that is widely used as a building block in instrumentation circuits; its schematic symbol is shown in Figure 5.8a. It differs from the simple amplifier of Figure 5.7 in that it has two inputs: one is an inverting input (designated by −), and the other is a noninverting input (designated by +). As suggested by the terminology, an increase in the voltage at the inverting input causes a decrease in the

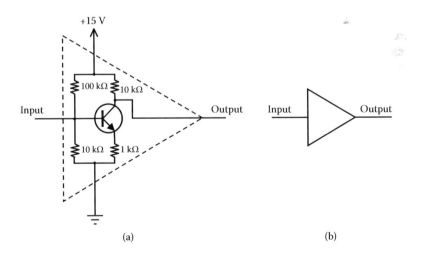

(a)   (b)

FIGURE 5.7 Amplifiers: (a) a simple amplifier circuit, consisting of one transistor and four resistors. (b) the schematic symbol for a generic amplifier.

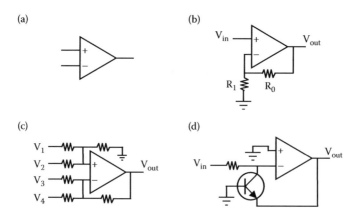

FIGURE 5.8 Operational amplifiers: (a) the schematic symbol for an op-amp, showing the non-inverting (+) and inverting (–) inputs; (b) negative feedback in an op-amp circuit; (c) an op-amp in a differential amplifier configuration; (d) a simple logarithmic amplifier consisting of a bipolar transistor in the feedback loop of an op-amp.

output voltage, but increases at the noninverting input cause the output to increase. Any difference between the two inputs is amplified by a factor of more than 10^5, so the circuit must be stabilized by connecting the output to the inverting input. Figure 5.8b shows resistor R_0 making this connection; reactive components (a capacitor or inductor) can also be used to limit the frequency response of the circuit. This negative feedback provides stability to the circuit since any voltage fluctuations at the output are immediately countered in response to the corresponding fluctuation at the inverting input.[6]

An op-amp has two defining characteristics: the inputs have extremely high impedance (and therefore draw no current), and the feedback loop ensures that the output voltage changes to maintain zero potential difference between the two input terminals. These two features explain the operation of the circuit shown in Figure 5.8b. The op-amp adjusts V_{out} so that the voltage at the inverting input is the same as the input voltage V_{in} at the noninverting input. From equation 5.19, this voltage is equal to the current I times R_1. Since virtually no current flows in or out of the inverting input, the same amount of current must flow through R_0, and the total voltage drop across both resistors, $I(R_0 + R_1)$, must equal V_{out}. Therefore it follows that

$$V_{in} = I(R_1) = \left(\frac{V_{out}}{R_0 + R_1} \right)(R_1) \tag{5.34}$$

$$V_{out} = \left(\frac{R_0 + R_1}{R_1} \right) V_{in} \tag{5.35}$$

so the circuit is a noninverting amplifier with a gain factor greater than one.

A similar analysis shows that the circuit of Figure 5.8c is a differential amplifier. If all of the resistors have the same value, then the output voltage is

$$V_{out} = V_1 + V_2 - V_3 - V_4 \qquad (5.36)$$

This result demonstrates the use of operational amplifiers in analog computation. If numbers are represented by voltages, then the circuit shown in Figure 5.8c can calculate the difference between $(V_1 + V_2)$ and $(V_3 + V_4)$ instantly. Other mathematical operations, including multiplication, division, calculation of ratio, exponentiation, integration, and differentiation can also be implemented with op-amp circuits. Thus, many of the calculations associated with measurement techniques can be accomplished prior to digitization of the sensor's output.

A final op-amp example is the logarithmic amplifier, which is used to calculate the ratio of light intensities in the near-infrared sensor described in chapter 11. The circuit in Figure 5.8d shows a transistor in a feedback loop. The voltage drop from the base to the emitter in a transistor depends logarithmically on the amount of current flowing into the collector (Horowitz & Hill, 1980, p. 117). The collector current in this case is equal to the input voltage divided by the input resistor, and the output voltage is equal to the voltage drop between base and collector. Therefore

$$V_{out} = A \ln\left(\frac{V_{in}}{B}\right) \qquad (5.37)$$

where A and B are constants. Commercially available designs are more sophisticated than Figure 5.8d, but they use this basic concept.

5.4 Digitization

The discussion thus far has centered on analog circuits, the output of which can assume continuous values. A shortcoming of analog circuits is that they are susceptible to electrical noise, whether it is introduced via the power supply, the environment, or the input signal. A related drawback in terms of measurement is the limited precision that can be obtained with analog signals. For example, if electrical noise introduces a 1 mV (peak to peak) ripple in a 1 V output signal, then the precision of the output is only 0.1%.

Digital circuits are by design less prone to the effects of electrical noise, primarily because digital signals have only two distinct states: on or off (equivalently, true or false). These binary states are represented by two well-separated voltage levels, so a modest amount of noise has no effect on them. In order to represent integer numbers other than 0 and 1, many binary digits (bits) must be combined to make a binary number (see Table 5.1). Fractional numbers are represented by using a standard format of a 32-bit or 64-bit binary number that contains the *mantissa*, *exponent*, and *sign* of the value (IEEE Standard No. 754-1985). As in the decimal number system, the level of precision

TABLE 5.1 Equivalence of Various Integral Numbers in Binary, Decimal, and Hexadecimal Format

Binary	Decimal	Hexadecimal
$2^7\ 2^6\ 2^5\ 2^4\ 2^3\ 2^2\ 2^1\ 2^0$	$10^2\ 10^1\ 10^0$	$16^1\ 16^0$
0 0 0 0 0 0 0 1	0 0 1	0 1
0 0 0 0 0 0 1 0	0 0 2	0 2
0 0 0 0 0 0 1 1	0 0 3	0 3
0 0 0 0 0 1 0 0	0 0 4	0 4
0 0 0 0 0 1 0 1	0 0 5	0 5
0 0 0 0 1 0 1 0	0 1 0	0 A
0 0 0 0 1 1 0 0	0 1 2	0 C
1 0 0 0 0 0 1 0	1 3 0	8 2
1 1 0 0 0 1 0 1	1 9 7	C 5
0 1 0 1 1 1 0 0	0 9 2	5 C
1 1 1 1 1 1 1 1	2 5 5	F F

is limited only by the number of available digits, so analog values can be represented to an arbitrary level of precision by using more bits. The transfer of binary data is essentially flawless, because error detection and correction schemes exist in hardware and in software. Therefore modern process sensors use digital data acquisition and communication techniques, and the old 4–20 mA current loop control systems are being replaced by intrinsically safe fiber optic systems that are based on digital circuitry.

Since most physical observables are analog (rather than digital) quantities, at some point in the measurement process an analog voltage or current must be converted into its binary form. This conversion is performed by a digitizer (analog-to-digital converter, or ADC), whose performance is specified in terms of its input voltage range, the number of bits N used to approximate the analog value, and the digitization rate (s^{-1}) at which it can operate. Digitizers are used extensively in audio and video systems, computer input devices, and telecommunications. Digitizers are also found in any application that requires a computer to react to its environment. A closely related device is the digital-to-analog converter (DAC), which provides an analog output voltage.

Two basic approaches to digitizing an analog number are *flash conversion* and *successive approximation*. The fastest and most expensive approach is flash conversion, which compares the incoming analog signal with all 2^N possible voltage levels simultaneously, chooses the best match, and outputs the fixed binary code associated with the corresponding voltage level. A simple three-bit flash digitizer is depicted in Figure 5.9a, where eight comparator circuits simultaneously compare the input voltage to eight equally-spaced levels within the 0 V to 1 V range of the converter. These comparators are essential op-amp circuits that change the sign of their output when the input voltage exceeds the reference voltage. The circuitry identifies the highest fixed value that is less than the input voltage and sends that code to the digitizer's output. Note that the number of comparators required is 2^N, so most true flash digitizers have a resolution of eight bits or lower.

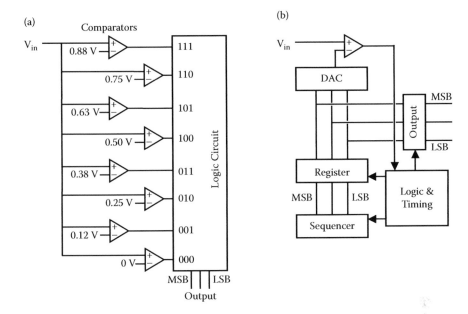

FIGURE 5.9 Analog-to-digital converters: (a) a flash digitizer; (b) a successive approximation digitizer.

Digitizers that operate by successive approximation include a DAC, a binary number register, and a comparator circuit (Figure 5.9b). Starting with the most significant bit, the circuit toggles each of the register bits in turn and notes whether or not the test voltage exceeds the input analog voltage. If the input is not exceeded, that bit is left on; otherwise, it is turned back off. After the least significant bit has been toggled, the register contains the binary equivalent of the input voltage level. This circuit is more complicated than flash digitizers, but the complexity is nearly independent of N, so much higher resolutions can be achieved (12-bit or 16-bit). The circuit is slower than the flash approach because it requires a sequence of comparisons; fortunately the conversion time is proportional to N, so resolution can be improved with only a modest decrease in digitization rate.

In practice, ADCs use a combination of flash and successive approximation techniques that is optimized for a specific application. In one representative design, a three-bit flash ADC quickly identifies the right voltage range, and two four-bit flash units are used in succession. This approach yields a 12-bit digitizer that operates at a rate of 1.25 megahertz (MHz).[7] Much higher resolution and speed are available.[8]

Suggested Reading

Halliday, D., Resnick, R., and Walker, J. (2005). *Fundamentals of Physics*, Seventh Edition (chaps. 21-30). New York: Wiley.

Horowitz, P. and Hill, W. (1980). *The Art of Electronics*. New York: Cambridge University Press.

Purcell, E.M. (1985). *Electricity and Magnetism*, Second Edition. New York: McGraw-Hill.

Kittle, C. (2005). *Introduction to Solid State Physics*, Eighth Edition. Hoboken, NJ: Wiley.

Yariv, A. (1985). *Optical Electronics*, Third Edition (chap. 11). New York: Holt Rinehart and Winston.

6

Ionizing Radiation

6.1 Introduction

Due to their frequent portrayal in science fiction stories and other media, radioactivity and X-rays are two well-known, but sometimes confounded, examples of ionizing radiation. In general, ionizing radiation is any form of subatomic particle or electromagnetic wave that propagates with sufficient energy to ionize atoms. This radiation can be divided into two broad classes: energetic electromagnetic waves and energetic particles.

An atom is composed of a positively charged nucleus surrounded by a "cloud" of negatively charged electrons that orbit at a very high speed.[1] The atom is electrically neutral because it contains equal numbers of positive and negative charge. The electrons are attracted to the nucleus by the electric force, but this force is balanced by the centripetal force of their orbital motion; this classical description is depicted in Figure 6.1a. Classical mechanics does not adequately describe atomic physics, however, because the electron must be under constant acceleration in order to maintain its orbit, and an accelerated electrical charge radiates energy in the form of electromagnetic waves. These radiated waves would continuously remove orbital energy from the electron, which would quickly spiral into the nucleus (Sears & Zemansky, 1955, p. 887). The stability of atoms was finally explained by the development of quantum mechanics in the early twentieth century.

The quantum mechanical description of an atom is based on the potential energy of interaction between electron and nucleus. The wave equation is solved for this potential (the potential for hydrogen is depicted in Figure 6.1b), and the solution is the wave function, the squared amplitude of which is the probability of finding the electron at a given radius from the nucleus.[2] An interesting result of quantum theory is that stable electron orbits can exist without angular momentum. An example is the ground state wave function, shown as a dashed line in Figure 6.1b for the case of hydrogen. The ground state wave function fits inside the potential well and is analogous to a standing wave in an acoustic or optic resonator. A number of additional wave solutions can also fit inside the potential well, and these represent higher-energy states of the same electron. The energy level associated with each wave function is discrete, so transitions between energy levels

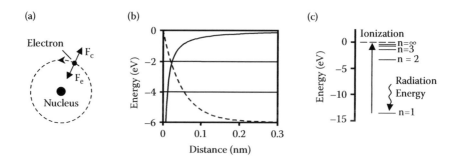

FIGURE 6.1 Atomic energy levels (a) a classical view of an electron orbiting a nucleus, where the centrifugal force F_c is balanced by the electric force of attraction F_e; (b) the potential energy between a proton and an electron is shown as a solid line; the ground state wave function for this system (i.e., the hydrogen atom) is shown as a dashed line; (c) energy levels and the ionization of hydrogen.

require quantized amounts of energy. In hydrogen, these energy levels (in electron volts) are given by

$$E_n = -\left(\frac{13.6}{n^2}\right)$$ (6.1)

where n is called the *principal quantum number*. Clearly the $n = 1$ state of the electron is the ground state because no electronic state exists at a lower energy. The energy of the ground state is -13.6 eV, which is stable because it lies below the minimum energy level needed to separate the electron from the proton (defined to be 0 eV). For illustration, several energy levels of hydrogen are depicted in Figure 6.1c.

Expressions similar to equation 6.1 can be written for the energy levels of other atoms and molecules. Since other atoms contain additional protons, they also contain additional electrons, and the electron-electron interaction modifies the energy levels. These electrons are arranged into shells corresponding to the principal quantum number. Inner-shell electrons require more energy to remove than ones on the outer shell of the atom. Other mechanisms such as spin-orbit coupling affect the energy levels in multi-electron atoms, but these details are not of immediate concern to us here.[3] The main point is that every electron in a given atom occupies a quantum mechanical state with a fairly well-defined energy that lies below the ionization limit.

Energy must be applied to an electron in order to free it from the atom. The energy that causes the transition from an atomic state to an ionized state is provided by some outside agency (which is called *ionizing radiation*). Ionization can be caused by a collision with another subatomic particle or by an interaction with a photon (i.e., an electromagnetic wave). A sketch of this process is shown in Figure 6.2a. When an atom loses an electron, a vacancy appears in the affected orbit; the atom's electrical charge is unbalanced, and the atom becomes a *cation* (a positively charged ion), which may be in an excited state.[4] If an ion has any remaining electrons, it can be ionized further; the subsequent ionization energies are of course much greater than the first, because there are fewer electrons to shield the positive charge of the nucleus.

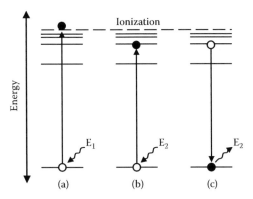

FIGURE 6.2 Ionization versus excitation (a) ionization ejects an electron and creates a vacancy; (b) excitation excites, but does not eject, an electron and creates a vacancy; (c) relaxation to the lower energy state produces characteristic electromagnetic radiation.

Under certain circumstances, an atom absorbs enough energy to promote one of its electrons to a higher (and unoccupied) energy level, but not enough to cause ionization (Figure 6.2b). This transition produces an excited (and therefore unstable) state of the atom, with a vacancy in the original orbit of the promoted electron.

Regardless of how electron vacancies are produced, they are quickly filled by one of the electrons from a higher orbital (or, in the case of ions, from the continuum of free electrons). Since this transition involves a change in energy level, the excess energy must be released, as depicted in Figure 6.2c. Transitions between bound electronic states (i.e., those below the ionization limit) give rise to electromagnetic radiation that is character-istic of the atom. This radiation is called *fluorescence*, and it is often used to identify the constituent elements in a sample.

6.2 Types of Ionizing Radiation

6.2.1 Energetic Electromagnetic Waves

X-rays and gamma rays are electromagnetic waves (as is light) that have wavelengths shorter than roughly 10 nm. In vacuum, all electromagnetic radiation travels at a speed of about 3×10^8 m/s and carries an amount of energy inversely proportional to the wave-length. Given an amount of energy E [eV], the wavelength of the light is given by

$$\lambda = \frac{hc}{E} \tag{6.2}$$

where h is Planck's constant (4.135×10^{-15} eV·s), and c is the speed of light in vacuum (3×10^8 m/s). Table 6.1 shows a few examples of this correspondence between electromagnetic wavelength and energy.

The distinction between X-rays and gamma rays is somewhat artificial and stems mainly from the source of the radiation as discussed below. X-rays typically have longer

TABLE 6.1 Correspondence Between
Electromagnetic Wavelength and Energy
(from Equation 6.2)

Energy	Wavelength (nm)
13.6 eV	91.2
50 eV	24.8
100 eV	12.4
500 eV	2.48
1 keV	1.240
2 keV	0.620
5 keV	0.248
10 keV	0.124
20 keV	0.0620
50 keV	0.0248
100 keV	0.0124
500 keV	0.00248
1 MeV	0.00124

wavelengths and are created by transitions between electronic states in atoms, as in Figure 6.2c. Gamma rays have higher energy and therefore shorter wavelength; they are created by nuclear transitions involving the protons and neutrons within the atomic nucleus, particularly during radioactive decay. Still more energetic rays are created by supernovae and other cosmic cataclysms, which give rise to cosmic rays.

The energy levels associated with the radiation used in process sensors tend to be relatively low (10 keV–512 keV); this radiation is provided by X-ray generators and sealed radioisotope sources. When these photons interact with an atom, they have sufficient energy to knock electrons from the atom, thereby ionizing it and losing energy in the process. Note that a single high energy photon can ionize several atoms in its passage through a target material.

6.2.2 Energetic Particles

The second class of ionizing radiation comprises *energetic particles*, especially *alpha* and *beta* particles. An alpha particle consists of two protons and two neutrons bound together; it essentially is a bare helium nucleus. Due to its relatively large mass and charge, an alpha particle interacts strongly with atoms in its path. It has a short range of a few centimeters in air and it cannot penetrate the outer layer of skin. Eventually it acquires two electrons and becomes a stable helium atom. Beta particles are either electrons (which are negatively charged) or positrons (which have the same mass as electrons but are positively charged). Positrons are antimatter, and when a positron collides with an ambient electron the two particles annihilate each other, producing two gamma rays.

Particles of mass m that are moving with a velocity v have a kinetic energy E equal to

$$E = \frac{1}{2}mv^2 \tag{6.3}$$

Equation 6.3 is the classical expression for kinetic energy, where the mass of the particle is assumed to be a constant regardless of speed. When the particle is moving at a high rate of speed (one-tenth the speed of light, c, or faster), a relativistic correction must be made for the mass:

$$m = \frac{m_0}{\sqrt{1 - \left(\dfrac{v}{c}\right)^2}} \tag{6.4}$$

where m_0 is the *rest mass*, or the mass of the particle measured in the frame of reference where the particle is not moving.

Ionization can occur if an energetic particle collides with a bound electron; some of the kinetic energy of the particle is transferred to the electron, which is classically speaking knocked out of its orbit. Ionization can also occur if the particle excites the nucleus to a higher energy state via a collision; when the nucleus decays to its ground state, energy is released that might cause autoionization of the atom. In all cases of ionization, the atom eventually decays back to its ground state.

6.3 Sources

6.3.1 Nuclear Decay

The nucleus of an atom is composed of protons and neutrons bound together in a very small volume on the order of 10 fm (i.e., 10^{-14} m) in diameter. A proton is the positively charged nucleus of a hydrogen atom, and a neutron is a charged particle with approximately the same mass as a proton. Each configuration of protons and neutrons in the nucleus has a specific energy associated with it in the same manner that electronic states have energies associated with them.

Radioactive nuclei are those that spontaneously decay to a lower energy state while emitting energetic particles (alpha and beta particles) and electromagnetic radiation (gamma rays and X-rays). After the transition, the nucleus has become a different isotope, which may or may not be stable; if it is unstable, it too will decay into another isotope until a stable nuclear configuration is reached. Due to the random nature of this process, a radioactive sample contains many different isotopes and related elements that are the products of the decay process.

The *activity* of a source refers to how frequently a decay event occurs; since each decay produces particles or rays, the activity is proportional to the intensity of the source. Activity is measured in the International System of Units as units of becquerel (Bq), which is one decay per second. The older unit was the curie (Ci), which equaled

3.7×10^{10} Bq. The energy of the rays produced by the radioisotope depends on the nuclear decay scheme and is characteristic of that radioisotope.

Radioactive materials can be produced in nuclear reactors, but they also occur naturally. A typical radioactive source used in process sensors is a small amount of a radioisotope that has been sealed inside a container. An external shield completely surrounds the container, except for a hole that allows the radiation to exit. A mechanical shutter over this hole provides a means of cutting off the radiation when necessary.

6.3.2   X-Ray Generators

An *X-ray generator* is essentially a vacuum tube in which an energetic beam of electrons is focused on a metal target. The electron source is an electrical filament that is heated by passing current through it. The filament becomes so hot that electrons are boiled into the vacuum, where they are extracted and accelerated by an electric field (Figure 6.3). Electrodes focus the electron beam by creating an electric field that acts as a lens, and the high-energy beam—typically 50 keV or more—slams into a metal target.

The kinetic energy of the electrons is released into the target, where a portion of it ionizes the electrons in the target, another portion creates high-energy photons directly, and the rest (which is most of it) appears as heat. The ionization of the target atoms creates vacancies in the electronic structure, and these vacancies (which represent a lower energy state) are quickly filled by other electrons in the metal target. In order for other electrons to fill these vacancies, they must liberate an amount of energy equal to the difference between their initial energy level and the energy level of the vacancy (as in Figure 6.2c). They liberate this energy by emitting a photon of the same energy, and so the wavelength is given by equation 6.2. These photons are fluorescent X-rays, and they are emitted at characteristic wavelengths that depend on the beam energy and the type of atoms in the target.

When the electrons slam into the target, they undergo significant deceleration. A free charge that is accelerated (or in this case, decelerated) can emit radiation as stated before, so some of the beam's kinetic energy goes into the creation of photons. This process is

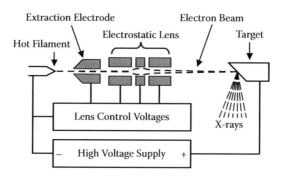

FIGURE 6.3 An X-ray generator.

called *bremsstrahlung* (from the German for "braking radiation"). Unlike the fluorescence X-rays produce by excitation of discrete energy states in the target atoms, bremsstrahlung photons have a continuous energy spectrum.

The X-rays caused by fluorescence and bremsstrahlung originate within a fairly small region of the target. The *spot size* of an X-ray generator is determined by the focus of the electron beam. In imaging applications, the spot size determines the resolution of the radiographic image, so vendors try to minimize it.[5] Putting so much energy into a small region can pit the target and reduce the ability to maintain the focus, so the spot size is limited by the local energy density that the target can withstand.

The heat produced by the impact of the beam on the target must be removed. Many industrial and medical X-ray tubes are water-cooled, and the water is pumped through a heat exchanger to cool it. High-power X-ray generator use a rotating target that spins at a high rate of speed in order to distribute the heat over a wider area.

6.3.3   Cosmic Rays

Equation 6.4 shows that mass (and therefore kinetic energy) increases significantly for particles that are moving at speeds approaching that of light. Supernovae, black holes, and other cosmic processes release inconceivable amounts of energy by converting some of their mass to energy. The concept that mass and energy are equivalent is expressed by Albert Einstein's famous equation

$$E = mc^2 \tag{6.5}$$

This energy is available to accelerate various subatomic particles (atoms and molecules being unable to survive under such conditions) to near-light speeds. Some of this energy also appears as very high energy photons.

When these highly energetic particles and photons reach the earth's upper atmosphere, they interact with the molecules there and create showers of secondary particles overhead. All of these particles and photons collectively are cosmic rays, and they contribute to the background radiation level. Cosmic rays are often a source of unwanted counts in radiation-based process measurements.

6.4   Detectors

6.4.1   Counters

The detection of X-rays and gamma rays is based on their ability to ionize matter. Counters are detectors that use ionization events to record the passage or energy of radiation. A well-known radiation counter is the *Geiger counter*, which detects radiation that passes through its thin-walled metal tube (Figure 6.4a). The Geiger counter uses a high voltage source to maintain a potential difference of several thousand volts between a center electrode and the wall of the tube, which is filled with an inert gas. When a gamma ray or X-ray passes through the tube, it partially ionizes the gas; the resulting ions accelerate due

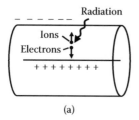 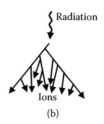

FIGURE 6.4 The Geiger counter: (a) a Geiger counter tube; (b) an avalanche of ionization that occurs within the Geiger counter tube.

to the strong electric field in the tube and create additional ions by colliding with other gas atoms. Within a very short time, secondary and tertiary collisions lead to an avalanche of ions and electrons racing toward the electrodes (Figure 6.4b). This avalanche is a current pulse that, when fed to a speaker, produces the familiar clicking sound of a Geiger counter. Versions of the Geiger counter can also detect high-energy particles, such as alpha rays.

Normally the avalanche produced in the tube gives a pulse that is independent of the energy of the incident radiation, but under certain conditions the pulse is proportional in height to the energy. This type of gas-filled detector is called a *proportional counter*.

Another type of counter is the *scintillation detector*. When radiation passes through clear materials, a flash of light is emitted as the ionized atoms recombine with electrons and fluoresce.[6] The intensity of the light (the number of photons) is proportional to the energy of the gamma ray or X-ray that passed through the scintillator. The flash of light is usually weak, so a very sensitive light detector is needed. Most detectors rely on the photomultiplier tube to convert these photons into a measurable signal; the output of the photomultiplier is a current pulse whose height is proportional to the intensity of the light and therefore to the energy of the original radiation.

Semiconductors can also be used to detect radiation. When high energy rays pass through the semiconductor, pairs of electrons and holes are created by ionization. If an electric field is applied across the semiconductor, the electrons and holes flow toward the opposite electrodes, thereby conducting current. The mechanism of radiation detection is therefore the same as in the case of visible light detection. As in the scintillation detector, the number of electron-hole pairs that are created is proportional to the energy of the incident ray. Due to the high energy of the radiation, thousands of charge carrier pairs can be created for each ray, and the resulting current pulse is analyzed according to height as in the case of scintillation detectors.

6.4.2   Imaging Devices

Detectors that are used for imaging applications, such as radiography and X-ray computed tomography, must preserve the spatial variations of radiation intensity that convey the X-ray image. In these applications measuring the radiation energy is of less importance than recording the radiation intensity. Imaging detectors span the range from radiographic film to the solid-state devices used in airport security screening to produce an X-ray image of passengers' luggage.

Photographic film has been used as a detector of radiation since the discovery of X-ray radiography in the late 1800s. When electromagnetic radiation strikes the silver halide salts in a photographic film, a chemical reaction takes place that creates a dark spot on the film after it is developed with the necessary chemicals. The grain size of the silver halide crystals determines the resolution of the film, and the intensity of the (visible or X-ray) radiation impinging on the film determines how dark that spot will be after the film is developed. The darkness, or gray level, of the film is therefore a record of the radiation intensity at that spot.

Since X-rays are able to pass through matter, very little of their energy is actually deposited in the film. Scintillation screens are often used in conjunction with films to increase their sensitivity. The X-rays pass through the film and strike the scintillation screen (also called an *intensifying screen*), which gives off visible fluorescent light in response to the radiation. This light is detected by the film, which is placed in intimate contact with the screen. After exposure, the film is developed as before; the image is substantially the same as a radiograph made without an intensifying screen, except that a lower dose of radiation (see below) is required. Such screens are often used in conjunction with medical applications of radiography. The disadvantage of intensifying screens is that they reduce the resolution of the image due to scattering of light within the screen itself.

In the early days of radiography, fluorescent screens were used to view X-ray images directly without the use of photographic film. The screens provided a glowing image of the internal structure of an opaque body; however, the dose of radiation was generally quite high, and it continued to increase as long as the object was under examination. This drawback may be acceptable in certain industrial applications, where the imaging can be viewed using a video camera, but it is clearly not appropriate for medical applications.

A related type of imaging method uses a fluorescent screen coupled directly with an *image intensifier* to produce real-time images at very low dosage. A similar but much more compact device is used for infrared imaging (so-called night vision). The basic X-ray image intensifier is depicted in Figure 6.5. These devices are large evacuated glass envelopes that are typically 20 cm or more in diameter; the input side of the intensifier is rounded in order to withstand the atmospheric pressure.

Referring to Figure 6.5, when radiation enters the intensifier it strikes a fluorescent screen and the photons emitted by that screen strike a photocathode that ejects electrons. The intensity pattern of these photoelectrons mimics the radiation field's original spatial pattern of intensity at the fluorescent screen. A high voltage of 20–60 kV is applied between the photocathode and a small phosphor screen at the output of the intensifier, and the resulting electric field causes the photoelectrons to accelerate. An electron lens, which consists of a series of apertures held at intermediate voltages, literally focuses the electron pattern onto the face of the phosphor screen. The energy of the electrons is therefore quite high by the time they strike the screen, so a single electron can excite thousands of phosphor molecules; this amplification is the origin of the image intensification. The excited phosphor fluoresces with visible light, which is visible through a small window at the exit of the intensifier. Due to the focusing of the photoelectrons, the image produced at the output is a small but visible and brighter image of the invisible X-ray pattern that entered the intensifier. This image can be viewed via an external video camera that is mounted to the image intensifier.

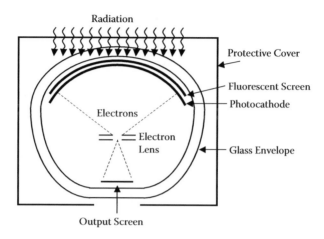

Radiation

Protective Cover

Fluorescent Screen

Photocathode

Electrons

Electron
Lens

Glass Envelope

Output Screen

FIGURE 6.5 Operation of an X-ray image intensifier.

X-ray image intensifiers are used in a variety of industrial applications that require real-time radiographic imaging, and an example is given in chapter 11. It can be noted here that the construction of large X-ray image intensifiers creates certain artifacts in the final image. The most obtrusive effect is the vignetting or tunnel vision effect caused by the curvature of the glass envelope. From Figure 6.5, it should be evident that X-rays near the edge of the intensifier actually travel through more glass than those near the center; the longer path length reduces the relative amount of radiation, causing the image to appear darker near the edges. This effect significantly skews the results of industrial measurements that are based on X-ray imaging, so the resulting real-time image must be normalized before it can be used (Scott, 1989).

Normalization of an image is a mathematical operation that maps the gray level L_0 (the relative brightness) at each pixel (x,y) in the image to a new value L_{norm}:

$$L_{norm}(x, y) = L_0(x, y) \cdot G(x, y) + B(x, y) \tag{6.6}$$

Here G and B are position-dependent values that correspond to a gain and baseline value for each pixel in the image. The normalization function described by equation 6.6 corrects for sensitivity variations in the intensifier as well as in the camera itself. The values of G and B are chosen so that the normalized gray level at each pixel meets two criteria: it equals 0 when the X-ray source is off (the baseline image), and it equals the maximum value (L_{max}) when the X-ray source directly illuminates the intensifier (the white image).[7] These criteria can be expressed as

$$0 = L_0'(x, y) \cdot G(x, y) + B(x, y) \tag{6.7}$$

$$L_{max} = L_0''(x, y) \cdot G(x, y) + B(x, y) \tag{6.8}$$

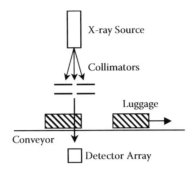

FIGURE 6.6 Use of a linear X-ray detector to screen luggage in airport security applications.

where L_0' is the intensifier's baseline image and L_0'' is the intensifier's white image. Equations 6.7 and 6.8 can be solved simultaneously for the values of G and B at each pixel. The requirement that the white image be uniform forces the value of G for image pixels near the edge of the intensifier to be larger than those near the center, in order to compensate for the vignetting effect noted above.

When an object is placed between the source and the intensifier, it casts a shadow on the input screen, and so the pixels in the normalized image have a gray value between 0 and L_{max} given by equation 6.6. This image can be used for quantitative measurements, as described in chapter 11.

The viewing area of image intensifiers tends to be limited by the size of the glass envelope, which must be able to support atmospheric pressure, and by the field of view of the electron lens, which tends to have a long focal length. An alternative approach to generating real-time or near real-time X-ray images is to use a mechanical scanner in conjunction with an array of solid-state detectors or scintillation detectors. A typical system is shown in Figure 6.6, where a narrow beam of radiation illuminates a closely-spaced linear array of detectors, which generates one row of pixels at a time. As objects pass between the source and detector, an X-ray image can be built from these rows. An advantage of this approach is that it reduces both dosage and X-ray scatter within the object (which degrades the radiographic image). Such devices are widely used in airport security operations.

6.5 Safety Considerations

The use of ionizing radiation in process sensors requires special consideration, since the improper use of it can cause bodily injury or death. The danger stems from the fact that ionizing radiation deposits enough energy in body tissue to damage or kill living cells. Large amounts of radiation received over a short time span can cause extensive tissue damage and even result in death, but such incidents are rare. Acute localized exposure to a limb or digit can result in the loss of the appendage. In most cases of exposure, however, there may be no immediate effect. Health effects from low or moderate doses of radiation include an elevated risk of developing leukemia or cancer, but these conditions often do not appear until several years after the exposure.

Radiation protection mainly consists of safety controls, such as adequate shielding and interlocks that contain or terminate the radiation whenever the shielding is opened. Depending on their energy, X-rays can usually be shielded by sheets of lead or even common steel. The X-ray generator itself is not radioactive, so these systems pose no radiation threat after they are de-energized. On the other hand, radioisotopes can only be shielded or contained, not deactivated. The gamma rays emitted by such sources are much more energetic than X-rays and require thicker shielding.

Of course, humans are constantly exposed to low-level radiation coming from a variety of natural sources, including cosmic rays and radioactive elements found naturally in rock. Radioactive decay of uranium produces radon gas, which can accumulate in houses and cause lung cancer through prolonged exposure. Therefore, the purpose of radiation shielding is to reduce the exposure to background levels, if possible.

The use of ionizing radiation in the workplace is regulated by the Federal Occupational Safety and Health Administration under Title 29 of the Code of Federal Regulations. These regulations stipulate that employers may not use or transport sources of ionizing radiation that cause anyone to receive a dosage in excess of specific limits during any one calendar quarter (U.S. Code of Federal Regulations, Title 29, Part 1910.1096). Sealed radioactive sources must be periodically checked for leaks and inventoried; generators and sources are also regulated by the states and other federal agencies.

The monitoring and reporting requirements associated with maintaining radiation sources is rather stringent and therefore costly. Proper disposal of radioactive sources is very expensive. For these reasons, many vendors that use radioactive sources in their sensors now offer very low-activity (less than 10^5 Bq), low-energy sources similar to those found in smoke detectors. Such sources are deemed to pose no significant threat; they are provided under a license held by the vendor, so the user is exempt from the more onerous regulatory requirements associated with larger radioactive sources. This is an attractive option if a weak source is sufficient for the application.

Suggested Reading

Griffiths, D.J. (1995). *Introduction to Quantum Mechanics*. Englewood Cliffs, NJ: Prentice Hall.

Winter, R.G. (1986). *Quantum Physics*, Second Edition. Colorado Springs, CO: IPI Press.

7

Conventional Sensors

7.1 Introduction

Conventional sensors are defined here to be those that meet three criteria: they must measure a single scalar quantity at a single point in the process, utilize technology that has been in widespread plant use for over two decades, and be available from a variety of commercial vendors. In these respects they differ from the more advanced sensors described in subsequent chapters, but both conventional and advanced sensors rely on the same underlying physical phenomena to make process measurements.

There are literally thousands of different sensor designs commercially available to monitor and control industrial processes, so it is impossible to present a comprehensive list of them here. Instead, several different technologies representing common process sensors will be presented for each type measurement. Additional information about conventional process sensors can be found in the books listed in the suggested reading for this chapter.

7.2 Temperature Sensors

Temperature is a basic measurement used throughout many processes. Many American plants still use the traditional Fahrenheit scale of temperature, but the Celsius (originally called centigrade) scale is the standard unit of measurement. Conversions between the two scales are given by

$$F = \left(\frac{9}{5}\right)C + 32 \tag{7.1}$$

$$C = \left(\frac{5}{9}\right)(F - 32) \tag{7.2}$$

Temperature is also expressed in terms of the Kelvin scale, whose zero point is defined to be −273.15 °C, which is absolute zero. The amount of heat contained in a body at a temperature of 0 °K is zero, and thermodynamic energy is calculated on this scale.[1] A temperature difference of one degree is the same in both the Kelvin and Celsius scales, so the conversion formula is

$$K = C + 273.15 \tag{7.3}$$

7.2.1   Thermistors

The word *thermistor* comes from a contraction of *thermal resistor*. The resistance of a thermistor is a function of the ambient temperature. To a first-order approximation, over limited temperature ranges the change in resistance ΔR of the thermistor is proportional to the change in temperature ΔT:

$$\Delta R = \alpha \Delta T \tag{7.4}$$

where α is the characteristic temperature coefficient of the thermistor. The temperature coefficient, which has units of $\Omega/°C$, can be either positive or negative, depending on the composition of the specific thermistor.

Positive thermal coefficient (PTC) thermistors increase their resistance as the temperature increases. A common type of PTC thermistor is formed by dispersing finely divided conductive particles in a thermoplastic or thermosetting matrix (Kohler, 1966). The matrix contains a sufficient amount of the conductive particles to form percolation chains (i.e., unbroken chains of particles) that conduct current through the polymer. As the temperature increases the volume expansion of the polymer begins to separate some of the links in those chains, with the result that the resistivity of the material increases. Another type of PTC thermistor is a barium titanate ceramic with small amount of rare earth ions such as bismuth or antimony (Matsuoka et al., 1976). Such materials typically have a limited temperature range over which they exhibit a positive temperature coefficient.

Negative thermal coefficient (NTC) thermistors reduce their resistance as the temperature increases. Many NTC thermistors are formed of semiconductor material, which has a band gap between the valence and conduction bands (Figure 5.1). The number of electrons in the conduction band is determined by the temperature and the Fermi level E_F of the semiconductor, which is the energy level occupied by the electrons at a temperature of absolute zero (0 °K). The probability $p(E)$ that an electron has an energy E is given by

$$p(E) = \frac{1}{e^{(E-E_F)/kT} + 1} \tag{7.5}$$

where k is Boltzmann's constant and T is the temperature [°K] (see Yariv, 1985, p. 471). It is clear that as the temperature increases, the energy distribution is skewed toward higher energies, and more electrons are promoted to the conduction band. It follows from equation 5.12 that an increase in the number density of charge carriers allows a larger current to be carried for a given potential drop. From the definition of resistance in equation 5.14, it then follows that the resistance of the thermistor decreases with increasing temperature.

Process temperature is measured with a thermistor by passing a constant current I through it, as shown in Figure 7.1a. The voltage drop across the thermistor is then the product of the current and the thermistor's resistance. From equations 5.19 and 7.4, it can be shown that the temperature T can be determined from the observed voltage V by

$$T = T_0 + \frac{(V - V_0)}{I\alpha} \tag{7.6}$$

where V_0 is the voltage drop measured at temperature T_0. Any error in the measurement of the voltage will obviously translate into an error in measured temperature. Another potential source of error is the constant current source.

Since a current flows through the thermistor, a small amount of heat is generated. The extra heat will raise the temperature of the thermistor slightly, producing an over-estimate of the process temperature. The power dissipated by a resistor is the resistance times the square of the current, so this effect can be minimized by keeping the current through the thermistor very small.

Due to the finite heat capacity of the body of the sensor, a thermistor does not respond instantaneously to changes in temperature. In fact, the time rate of change of the measured temperature (resulting from an abrupt increase in process temperature) can be defined in terms of a time constant τ so that

$$\frac{dT}{dt} = \frac{(T - T_0)}{\tau} \tag{7.7}$$

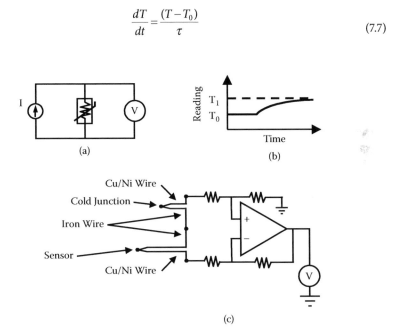

(a)

(b)

(c)

FIGURE 7.1 Temperature sensors: (a) measurement of the voltage drop across a thermistor; (b) time lag between temperature increase and sensor response; (c) measurement of the voltage produced by a type J thermocouple using an operational amplifier and a cold junction.

This differential equation can be solved for the measured temperature $T(t)$, which is plotted in Figure 7.1b as a function of time:

$$T(t) = (T_1 - T_0)[1 - e^{-t/\tau}]$$ (7.8)

where T_0 is the initial temperature and T_1 is the final temperature. Thus, the response time of the thermistor or any other temperature sensor depends on the value of τ, which is a function of the mass of the sensor. Fast response times are achieved by minimizing the size and weight of the sensor. The issue of response speed is not critical for slow reactions or stable processes, but in any case the characteristic time τ of the sensor should be short compared to the time scale of the process.

7.2.2   Thermocouples

A *thermocouple* is commonly used to measure the temperature in a chemical process because it is inexpensive, rugged, and small. It is ideal for detecting rapid changes in temperature, since they have little mass and therefore a very small heat capacity.

The operation of thermocouples is based on the discovery, made in the 1820s by Thomas Seebeck, that a circuit formed from wires of two dissimilar metals generates an electric current when one of the junctions is heated. This current is due to an electric potential that develops across the hot junction; the potential difference (ΔV) depends only on the composition of the two wires and the temperature difference (ΔT) between the two junctions:

$$\Delta V = \beta \Delta T$$ (7.9)

where β, measured in $V/°C$, is the characteristic thermoelectric coefficient for that combination of metals. The temperature at the hot junction can therefore be determined by measuring the thermoelectric potential.

The Seebeck effect is due to a difference in the density of conduction electrons between the two metals (Sears & Zamansky, 1955, p. 547). As mentioned above, the probability of finding an electron in the conduction band depends on temperature and the characteristic Fermi level of the metal. The difference in Fermi level is bound to leave one metal with more conduction band electrons than the other. When the two metals come into contact with each other, the relative excess of free charges moves across the junction from one to the other metal. This movement of charge constitutes an electric current. As the temperature of the junction increases, two effects happen: the electrons have a higher average velocity, and the disparity in density of the conduction electrons becomes more pronounced. Therefore the potential difference increases with temperature.

The voltage produced by the thermocouple is measured with an operational amplifier or other instrumentation amplifier, as shown in Figure 7.1c. A second thermocouple (the cold junction) is held at a known temperature to compensate for the offset voltage produced by the thermoelectric effect at the connection between the thermocouple wire and the instrument (Horowitz & Hill, 1980, p. 592). Traditionally, the cold junction was placed in an ice water bath, which has a constant temperature of 0 °C, but modern sensors compensate for this offset voltage electronically. The observed voltage V is converted to temperature using the formula

$$T = \frac{(V - V_0)}{\beta} + T_0 \tag{7.10}$$

This conversion is often provided automatically by the sensor electronics.

It is important to note that thermocouples measure the temperature difference between two points, not absolute temperature. Various metal alloys have been developed to maximize the thermoelectric potential, and standard thermocouple junctions have been defined to simplify calibration and measurement. Type J thermocouples (shown in Figure 7.1c) consist of iron and copper-nickel wires and have a thermoelectric coefficient of 51 (μV/°K) at room temperature; they can be used up to 700 °C. Type K thermocouples consist of nickel-chromium and nickel-aluminum alloys and have a coefficient of 41 (μV/°K) at room temperature; they can be used at temperatures over 1000 °C. Thermocouples are normally protected by a sheath that is made of stainless steel or Inconel.

7.2.3 Infrared Thermometers

Objects that are at a higher temperature than that of their environment radiate electromagnetic radiation; this fact is demonstrated by the light emitted by glowing objects that are "red hot" or "white hot." The qualitative observation is that objects hotter than about 500 °C emit a dull red glow; as the temperature of the object increases, the light becomes brighter, and also whiter. This emission is called *blackbody radiation*, and it is caused by the thermal excitation of the electrons, which radiate energy. The electronic excitation and radiation increase with the temperature of an object.

Even objects that are slightly above room temperature emit light; it is infrared and therefore cannot be seen by eye, but with a suitable detector this light can be used to measure temperature without contacting the object. Noncontact measurements are especially useful when the temperature exceeds the useful range of thermocouples (for example, in steel-production furnaces).

The average thermal energy of particles at a temperature of T (measured on the absolute scale in units of °K) is on the order of kT, where the Boltzmann constant k = 8.617 × 10^{-5} eV/°K. At temperatures near 500 °C, this average energy is only a small fraction of the several electron-volts of energy necessary to create a photon of visible light, so the emission of such photons by a hot object may seem surprising. The explanation lies in the fact that thermal interactions are random, and the probability exists that at any given temperature, an electron might reach a much higher energy state and radiate this energy as an electromagnetic wave. The intensity distribution of blackbody radiant energy is actually a function of the electromagnetic frequency f (Feynman et al., 2006, vol. 1, Lect. 41):

$$I(f) = \frac{4hf^3}{c^2(e^{hf/kT} - 1)} \tag{7.11}$$

where h is Planck's constant and c is the speed of light. Since the energy of a photon is equal to the product (hf), the intensity distribution of equation 7.11 can be expressed in terms of energy:

$$I(E) = \frac{4E^3}{(hc)^2(e^{E/kT} - 1)} \tag{7.12}$$

The relative magnitude of this function is compared at several temperatures in Figure 7.2. It can be seen that energy of the radiation shifts upward as the temperature increases.

The total amount of power P (i.e., energy per unit time) emitted by a body is a function of its temperature above ambient (ΔT):

$$P = \sigma \varepsilon A (\Delta T)^4 \tag{7.13}$$

where σ is the Stefan-Boltzmann constant (5.67×10^{-8} W·m^{-2}·K^{-4}), ε is the emissivity of the surface, and A is the surface area of the object. The emissivity is a dimensionless number constant with a value between 0 and 1; it indicates the relative ability of a surface to absorb and emit electromagnetic radiation. By capturing a portion of the infrared light emitted by an object, one can measure the relative power of the radiation and therefore estimate the temperature difference ΔT from equation 7.13.

The basic infrared thermometer uses a lens to focus the blackbody radiation onto a cooled infrared detector. Best results are obtained with a cooled detector, because thermal noise is reduced at low temperatures. In many modern detectors, the cooling is provided by a Peltier junction, which uses the reverse thermoelectric effect to carry heat away from the bimetal junction. The detector itself can be a solid state device, such as those described in chapter 5. Another type of detector used in noncontacting thermal measurements is the *thermopile*, which is an assembly of 20–100 thermocouples wired

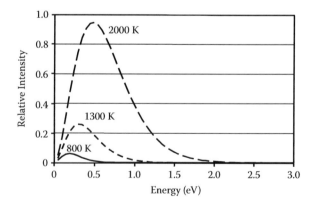

FIGURE 7.2 Relative intensity of black body radiation as a function of photon energy, shown at temperatures of 800 °K, 1300 °K, and 2000 °K.

in series that share a common hot junction (the active part of the sensor) and a common cold junction. When the radiation is absorbed at the hot junction, the energy it carries is released as heat, and the thermopile generates a small potential. The temperature of the thermopile is the result of a balance between the amount of heat being deposited by the absorbed radiation and the amount of heat being lost through conduction or reradiation. A steady state is quickly established (within a few characteristic times τ of the detector), and the resulting voltage from the thermopile indicates the relative energy being deposited by the blackbody radiation.

7.3 Pressure Sensors

Pressure is defined to be the net force exerted per unit area on a surface. In gases and liquids, this pressure may come from the weight of the fluid (at the bottom of a tank), or it may come from the kinetic energy associated with the thermal motion of the fluid molecules. The unit of pressure is the *pascal* (Pa), which is defined to be 1 newton per square meter. Pressure measurements in the plant are often expressed in terms of *gauge pressure*, which is the differential pressure in excess of the ambient or atmospheric pressure. Absolute pressure is expressed with respect to vacuum, and is therefore one atmosphere (0.1 MPa) more than gauge pressure.

The hydrostatic pressure at a depth d (in meters) in a liquid-filled tank is given by

$$p(d) = \rho g d + p_0 \tag{7.14}$$

where ρ is the density [kg/m^3] of the fluid, g is the standard acceleration due to the earth's gravity (9.8 m/s^2), and p_0 is the pressure of the gas in the head space above the liquid. The depth d is the vertical distance measured from the surface of the liquid to the sensor port.

In a gas, the pressure is determined by the free (gas-filled) volume V of the vessel, the absolute temperature T, and the number of molecules N in the gas. In an ideal gas, the molecules are assumed to have no interaction, and the product of pressure and volume is

$$PV = NkT \tag{7.15}$$

where k is again the Boltzmann constant. The ideal gas law breaks down at high pressures and low temperatures, where intermolecular forces begin to dominate the kinetic properties of the gas.

7.3.1 Diaphragms

Since pressure is by definition the force exerted on a unit of surface area, pressure can be measured directly by measuring the force on a disc of known area. A rigid disc may be used if it is mounted on bellows that allow movement in response to the force; in that case the force may be measured directly with a force sensor. A more common design uses a flexible diaphragm or membrane that deforms elastically under pressure. Once the deflection is measured, it can be converted to force (and ultimately to pressure) by using a mechanical model of how the diaphragm deforms under a uniform load.

A generic pressure gauge is shown in Figure 7.3. The diaphragm in the sensor of Figure 7.3a is exposed on both sides. The net force on the diaphragm is the difference in pressure between side A and side B, multiplied by the area of the diaphragm:

$$F = A(p_A - p_B) \tag{7.16}$$

The physical observable in this case is the deformation, which is measured using one of the devices described below. If the port on side A is connected to the process; and, the port on side B is left open to the atmosphere, then the sensor reports gauge pressure. The same sensor can also be used to measure pressure differentials by connecting ports A and B to different points in the process. If however side A is evacuated of air and sealed, as shown in Figure 7.3b, then the sensor measures absolute pressure.

When pressure is applied to the diaphragm, it deforms by an amount d shown in Figure 7.3c. The exact relationship between the applied pressure and diaphragm deformation depends on the details of the sensor design. For relatively low pressures, a thin membrane of radius r can be used, and the deformation is directly proportional to the pressure (Fraden, 2004, p. 340):

$$d = \frac{r^2 p}{4S} \tag{7.17}$$

Here, S is the radial tension applied to the membrane. Of course, a thin membrane is likely to burst at high pressures, so a thin plate can be used instead. Although the equation for the deformation is a bit more complicated than in the case of the thin membrane, it turns out that for small deformations, d is still proportional to the applied pressure p.

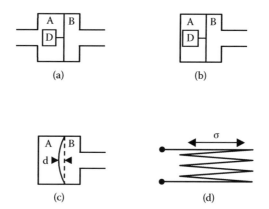

(a)   (b)

(c)   (d)

FIGURE 7.3 Pressure sensors: (a) a two-port diaphragm sensor for measuring differential pressure; (b) a one-port diaphragm sensor for measuring absolute pressure; (c) deformation of a diaphragm under pressure; (d) foil pattern for a strain gauge used to measure deformation in pressure sensors.

Therefore, pressure sensors are based on the measurement of the corresponding deformation in the membrane or diaphragm.

7.3.2   Capacitance Manometers

One way to measure the deformation of a conductive diaphragm is to treat it as one side of a parallel plate capacitor. Capacitance is inversely proportional to the distance between the two plates, so movement of the diaphragm causes an observable change in the electrical capacitance. One method of detecting this change is to build an electronic oscillator circuit whose frequency is determined by the capacitance of the diaphragm; such a device is called a *capacitance manometer*. Manometers are typically used to measure pressures less than one atmosphere.

7.3.3   Strain Gauges

The most common type of pressure sensor uses *strain gauges* to monitor the deformation of the diaphragm. A strain gauge is an electrical device whose output signal is proportional to the amount of tensile or compressive strain placed on it by outside forces. Strain is the relative or fractional change in length, so for example a rod of nominal length L that elongates under tension a distance ΔL experiences a strain σ, where

$$\sigma = \frac{\Delta L}{L} \tag{7.18}$$

If a strain gauge is glued securely to the rod, then the strain gauge will experience the same strain σ.

A simple strain gauge is based on the resistance of a conductive material such as metal foil. Since the resistance of a wire is proportional to its length, it follows that stretching a wire increases its resistance. Resistive strain gauges typically have a foil pattern similar to that shown in Figure 7.3d in order to increase the sensitivity to strain along the direction indicated. The strain σ is equal to the fractional change in resistance, so pressure can be measured by fixing a strain gauge to the diaphragm shown in Figure 7.3c and monitoring the resistance.

Modern strain gauges that are used in pressure sensors are made of silicon or other semiconductor material (Horowitz & Hill, 1980, p. 602). Such devices have two benefits over the older type of foil strain gauge: they have a much higher resistance and therefore can be read with a constant current source, and they can be built onto an integrated circuit as a complete circuit including an instrumentation amplifier and other support circuitry. This circuitry sends a constant current through the strain gauge and records the voltage that develops.

7.4   Level Sensors

Level sensors are used to monitor the amount of liquid or particulate matter in a tank or storage bin. Thus, the sensor output is a distance, measured from either the top or the bottom of the tank. One way to estimate the level of liquid in a tank is to measure the gauge pressure at the bottom of the tank. This pressure is seen in equation 7.14 to be proportional to the liquid level *d*. Other sensing methods are described below.

7.4.1   Capacitive Sensors

A *capacitive level sensor* measures the electrical capacitance between a probe electrode (illustrated in Figure 7.4) and a second electrode, which is often the metal wall of the tank. The capacitance of the center electrode changes as a function of height of the material, starting at its lowest value when the tank is empty, and reaching its maximum value when the tank is full. The actual value of the capacitance depends upon the geometry of the electrode and tank and the dielectric constant of the material.

This type of sensor can be used in a wide variety of liquids and solids. To a large extent it can tolerate buildup of material on the electrode, as it usually represents a very small fraction of the cross-sectional area of the tank. However, this technique does have limitations: highly conductive material will short out the electric field generated by the center electrode. If the material is a conductive liquid, the resulting current will cause heating, which may lead to undesirable effects. In the case of conductive solid particulates or fibers, the heating will be concentrated at the interfaces between the particles and may fuse them together.

An alternative approach for monitoring the level of conductive liquids is to use the electrode configuration of Figure 7.4 to measure the resistance instead of the capacitance. Relatively modest currents can be injected into the center electrode, and the voltage drop between it and the tank wall can be measured. In this case, the resistance will be at a maximum when the tank is empty and at a minimum when the tank is full.

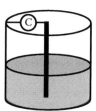

FIGURE 7.4 The level of material in a metal tank can be determined by measuring the capacitance *C* between a center electrode and the tank wall.

7.4.2   Ultrasonic and Acoustic Sensors

Another approach is to locate the interface between the top of the material and the gas in the head space. Since sound waves are reflected by a difference in density or acoustic impedance, the interface between the material and the head space generally provides enough contrast to create a reflection (echo). The frequency of the sound can be in the range of human hearing, or it can be ultrasonic, but in either case it must be a pulse of relatively short duration.

In the case of many liquids, this type of sensor can be mounted either in the bottom of the tank looking upward or in the top of the tank looking downward. In the case of powders or particulate matter, the sensor works best when mounted at the top of the tank. In either position, the sound travels until it encounters the interface, at which point an echo is formed. The echo travels back to the sound transducer, where it is detected. The propagation time (sometimes called the *time of flight*) of the sound in the liquid or gas is determined by the round trip distance to the interface and back. If the time delay between the emission of the pulse and the reception of its echo is denoted as τ, then the distance from the sensor to the interface is

$$d = \left(\frac{\tau}{2}\right)v \qquad (7.19)$$

where v is the velocity of sound.

The physical observable in this case is time, but error in the velocity of sound assumed in equation 7.19 may have a more detrimental effect on the process measurement than error in the time of flight. It turns out that sound velocity is generally very sensitive to temperature, so a potential problem with this approach is variation in temperature along the path of the sound. Even sensors that have a built-in temperature correction feature can not fully account for time of flight variations, as they measure the local temperature.

7.4.3   Radar Time-Domain Reflectometers

Radar can be used to make similar time of flight measurements even when there are temperature variations in the process vessel. In this case the technique is called *radar time-domain reflectometry*, but the basic concept is exactly the same as that used by the ultrasonic sensor. Miniature radar units generate pulses of electromagnetic radiation that can be scattered and reflected by materials, particularly if they are conductive. These sensors are mounted inside the vessel at the top, because the electromagnetic field cannot penetrate metal walls.

As in the case of the ultrasonic sensor, the time of flight τ is measured for the radar pulse. The velocity used in equation 7.19 is the speed of light, and it should be noted that the values of τ are quite small; typical values will be on the order of 5 to 100 ns. Therefore the measurement of time is much more critical in the case of radar than in the case of ultrasonic reflectometry (where typical time scales are 100 μs to 0.1 s). Very precise level measurements are therefore difficult to obtain using radar; better results can be obtained with an acoustic sensor.

7.5   Flow Rate Sensors

The *flow rate sensor* measures either the velocity v [m/s] or the mass flow rate $\dot{m}$ [kg/s] of fluids or fluidized particles in a pipe or process stream. If the average velocity v through a pipe is measured, then the mass flow rate can be estimated from

$$\dot{m} = A\rho v \qquad (7.20)$$

where A is the cross-section of the pipe [m^2] and ρ is the density of the material [kg·m^{-3}].

7.5.1   Heat Transfer Sensors

A common method of measuring flow in liquids or gases is to measure the amount of heat carried away by the process fluid. An example is the *hot-wire anemometer*, which comprises an electrically heated element (typically a hot wire or film) that is in thermal contact with the fluid stream to be measured. Thermal energy emanating from the wire is absorbed by the process fluid flowing past the sensor. Thus the temperature of the heated element is determined by the balance between heat flowing into the wire due to resistive heating and the heat flowing out of the wire into the process fluid. The flow rate of the fluid determines the amount of cooling provided to the sensor, so temperature is an indicator of flow rate, assuming the electrical current through the wire is constant.

Heat, represented as Q and measured in joules, is simply a transfer of thermal energy from one material to another. When heat flows to or from an object (in this case the flow sensor), the temperature of the object increases proportionally by an amount ΔT:

$$Q = K\Delta T \qquad (7.21)$$

Here, the proportionality constant K is called the heat capacity of the sensor. The heat Q represents the net energy gain. The process fluid carries heat away from the sensor and therefore limits both Q and the temperature rise.

A constant temperature anemometer generates just enough heat to maintain a constant temperature in the sensor; this temperature is slightly above the ambient temperature of the fluid. A control circuit in the sensor constantly monitors the temperature and adjusts the power input to the heater accordingly. One method of measuring the temperature of the wire is to measure its resistance, which has been shown earlier to be a function of temperature. In this type of sensor, the control circuit applies a short duration of heating current, followed by a resistance measurement of the hot element (Bernard & Collet, 2000). By alternating rapidly between the heating and measurement cycles, the sensor is able to maintain a constant temperature and simultaneously generate a signal that is proportional to the flow rate.

Of course, these sensors must be calibrated for the type of fluid that is contacting the sensor. The fluid carrying the excess heat away from the sensor has a specific heat K_m defined by

$$K_m = \frac{Q}{m\Delta T} \qquad (7.22)$$

where m is the mass of a given amount of fluid. Here the heat Q is the amount of heat needed to raise the temperature of a mass m of the fluid by an amount ΔT. Specific heat is a material property that varies from one fluid to another; it is also a function of temperature. For these reasons, thermal sensors do not work well in multiphase flow applications where the value of K_m may fluctuate.

7.5.2 Ultrasonic Sensors

As in the case of level measurement, the flow velocity in liquids or gases can be measured using an ultrasonic time of flight technique. Figure 7.5 shows a process pipe of diameter D with two ultrasonic transducers mounted in ports on opposite sides of the pipe. One transducer transmits pulses of ultrasound, and the other, which is mounted a known

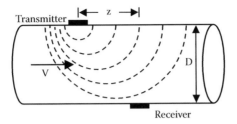

FIGURE 7.5 An ultrasonic flowmeter mounted in a pipe of diameter D. The velocity of the wavefronts of successive pulses (depicted by the dashed lines) increases downstream due to the velocity of the flow.

distance z downstream, receives the pulses. The frequency of the ultrasound is typically 1 MHz or lower for gas metering applications and 1–20 MHz for liquid metering applications. The ultrasonic frequency is chosen as a tradeoff between signal strength (which tends to be best at lower frequencies) and adequate pulse definition (which requires higher frequency content).

If the fluid velocity in the pipe were zero, then the time of flight of the ultrasound τ_0 would simply be the distance between the transducers divided by the speed of sound c_0 in that fluid:

$$\tau_0 = \frac{\sqrt{D^2 + z^2}}{c_0} \tag{7.23}$$

However, if the fluid is moving through the pipe with an average velocity v, then the sound waves will be carried along with the fluid as depicted in Figure 7.5, and the time of flight will be reduced. Of course, the flow velocity through a pipe is generally a function of position within the pipe (the velocity profile is parabolic for laminar flow), but the overall effect of the fluid flow will be to hasten the arrival of the pulse.

Using the approximation that the average velocity of flow in the pipe is v (in m/s), a simple expression can be derived for the time of flight. Figure 7.6a shows the geometry

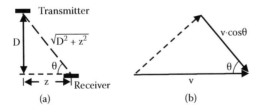

FIGURE 7.6 Geometry used in the calculations leading to equation 7.27: (a) the distance between the ultrasonic transmitter and receiver is the hypotenuse of the right triangle as shown, and the ultrasonic path is at an angle θ with respect to the pipe wall; (b) the flow velocity is resolved into two component vectors.

used in this calculation; given the relative positions of the transmitter and receiver, the sound must travel along a line that intersects the wall of the pipe at an angle θ, where

$$\cos\theta = \frac{z}{\sqrt{D^2 + z^2}} = \frac{z}{c_0 \tau_0} \tag{7.24}$$

As noted above, the sound waves are carried along with the flow of the fluid, so the effective sound velocity in the direction of the flow is $(c_0 + v)$, whereas the effective sound velocity transverse to the flow is only c_0. Since velocity is a vector, the velocity of the process fluid can be resolved into two orthogonal components, which are depicted in Figure 7.6b. The component of flow velocity along the direction defined by the angle θ is $(v \cdot \cos \theta)$, so in this direction sound propagates at a speed of $(c_0 + v \cdot \cos \theta)$.

The transit time τ (time of flight) of the pulse from the transmitter to the receiver is therefore equal to the distance divided by the effective sound velocity in that direction. From equation 7.23 it is evident that

$$\sqrt{D^2 + z^2} = c_0 \tau_0 \tag{7.25}$$

so it follows that the transit time is

$$\tau = \frac{\sqrt{D^2 + z^2}}{c_0 + v \cdot \cos\theta} = \frac{c_0 \tau_0}{c_0 + [vz/(c_0 \tau_0)]} = \frac{(c_0 \tau_0)^2}{c_0^2 \tau_0 + vz} \tag{7.26}$$

where the cosine has been substituted from equation 7.24. Solving equation 7.26 for v, we find that

$$v = \frac{c_0^2 \tau_0^2 - c_0^2 \tau_0 \tau}{z\tau} = \left(\frac{c_0^2 \tau_0}{z}\right)\left(\frac{\tau_0}{\tau} - 1\right) \tag{7.27}$$

Therefore, the velocity of fluid in a pipe can be measured by the time of flight method without explicitly defining the pipe diameter, provided the zero-flow transit time τ_0 and the distance between the two transducers are known. It is also possible to calibrate the sensor by extrapolating τ_0 from the transit time τ observed at several flow velocities.

Much of the uncertainty in ultrasonic flow measurement is due to temperature effects. Sound velocity depends on temperature T, so it is imperative to know $c_0(T)$ as well as the temperature in the pipe where the flow measurement is made. These sensors generally have some sort of temperature compensation built into them; the more advanced models can model $c_0(T)$ based on a number of calibrations at various temperatures.

Another scheme for measuring flow velocity in liquids via ultrasound is to detect the frequency shift in the echoes from particles or bubbles entrained in the flow. If a beam of ultrasound from a transducer is injected into the process stream, a portion of the sound waves will be reflected by any particles or bubbles that happen to be present. The ultrasound can be either a continuous wave or a tone burst of relatively short duration. In the case of a continuous wave, the echoes can be received by a second transducer; if a tone burst is used, a single transducer can receive the echoes during the delay between transmissions.

In either case, the frequency of the returning wave will be Doppler shifted, as described in section 3.4.5.

The frequency shift Δf was calculated in equation 4.12 for light reflected by a moving mirror, but the result is the same for ultrasound, provided c is interpreted as the speed of sound. Taking f_0 as the frequency of the ultrasonic wave emitted by the transducer, then it follows from equation 4.12 that the frequency of the echo f is determined by the velocity of the liquid v:

$$f = f_0 + \frac{2v}{c} f_0 \tag{7.28}$$

Solving for v, we find that

$$v = \frac{c}{2}\left(\frac{f}{f_0} - 1\right) \tag{7.29}$$

where v is positive if the flow is directed toward the sensor and negative if it is directed away.

The derivation of equation 7.29 has assumed that the ultrasound propagates along the direction of the flow, and some sensors designs place the transducer directly in the middle of the process stream. That arrangement is not optimal, since it restricts the flow and creates the possibility of a plug of material forming around it. Most of the Doppler flowmeters inject and receive the ultrasound at an angle θ with respect to the pipe. In such a sensor, the velocity that is actually measured is the component of flow velocity along that direction; referring back to Figure 7.6b, that component is equal to ($v \cdot \cos \theta$). Therefore, a more general statement of equation 7.29 is the following:

$$v = \frac{c}{2\cos\theta}\left(\frac{f}{f_0} - 1\right) \tag{7.30}$$

As in the case of the time of flight measurements, the Doppler technique is sensitive to changes in temperature. Flowmeters generally include a means of monitoring temperature in order to provide the correct flow rate. One can also turn this deficiency into a benefit: if the flow rate is known, then changes in the time of flight or frequency can be used to estimate the temperature from $c_0(T)$. That approach is the basis for ultrasonic thermometers.

7.5.3 Electromagnetic Sensors

Electromagnetic flowmeters are briefly mentioned here because they are so widely used; their operation is described in more detail elsewhere.[2] They can be used to measure electrically conductive liquids and slurries, which include quite a lot of materials of interest in industry. The operating principle is based on the fact that conductive liquids contain a high concentration of positive and negative ions. As these ions are carried forward by the process flow, they can be deflected by an external magnet whose lines of magnetic flux are perpendicular to the pipe. The resulting force on the ions causes them to move perpendicularly to the directions of both the liquid flow and the magnetic lines of flux

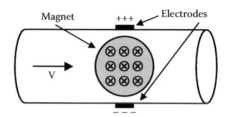

FIGURE 7.7 An electromagnetic flow sensor with its magnetic field directed into the page. Fluid flow is from left to right.

(Halliday et al., 2005, p. 737ff.). A generic sensor is shown in Figure 7.7, where the $\otimes$ symbol indicates that the magnetic flux is pointing into the page; if the flow is moving to the right as shown, then the positive ions will be deflected upward, and the negative ions will be deflected downward. If electrodes are placed in contact with the process fluid as shown, then an electric potential will develop between them, and the voltage will be proportional to the average flow velocity.

The magnetic field can be created by a permanent magnet or, more typically, by a pair of electromagnetic coils. The advantage of using electromagnets is that they can produce an alternating magnetic field, which in turn generates an alternating polarity on the electrodes. The observed voltage can then be measured using phase sensitive detection or other techniques that compensate for baseline drift in the output.

7.5.4   Differential Pressure Sensors

One of the most common flow sensors used in industry is the differential pressure gauge shown in Figure 7.8. Pumping either liquid or gas through a constriction, such as a smaller pipe diameter (Figure 7.8a) or an orifice (Figure 7.8b), creates a pressure drop across the constriction. This pressure differential increases with flow velocity, so the

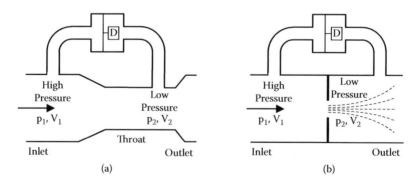

FIGURE 7.8 Measurement of flow using different pressure sensors in (a) Venturi meters and (b) Orifice plates. Fluid flow is from left to right.

pressure sensor of Figure 7.3a can be used to measure flow. A variety of geometries have been used to generate the pressure differential (Upp & LaNasa, 2002).

A typical sensor for measuring the flow of liquids is the *Venturi meter*, which is shown in Figure 7.8a. This device consists of a round tube that tapers from the inlet (on the left) down to a minimum diameter at the "throat" of the device and then expands at the outlet. The Venturi meter is designed to measure the flow velocity of an inviscid, incompressible fluid of density ρ. Its operation is based on Bernoulli's equation, which states that the pressure p and velocity v at any two points along a streamline in such a fluid (provided they are at the same elevation) are such that

$$p_1 + \frac{\rho v_1^2}{2} = p_2 + \frac{\rho v_2^2}{2} \tag{7.31}$$

Here, we are measuring the pressure differential $\Delta\rho$ between the inlet and the throat of the Venturi meter. As a first-order approximation, we assume that the velocity at the throat is proportional to the velocity at the inlet (Fraden, 2004, p. 358):

$$v_2 = K v_1 \tag{7.32}$$

where $K > 1$. It follows that

$$\Delta p \equiv p_1 - p_2 = \frac{\rho}{2}\left(v_2^2 - v_1^2\right) = \frac{\rho}{2}(K^2 - 1)v_1^2 \tag{7.33}$$

which means that the inlet velocity is proportional to the square root of the pressure drop:

$$v_2 \propto \sqrt{\Delta p} \tag{7.34}$$

In practice, the viscosity of the real fluid and the geometry of the Venturi meter affect the observed Δp, so the proportionality constant implicit in equation 7.34 must be found experimentally or estimated from engineering handbooks for a given application. Temperature effects must also be considered in the final calculation of flow velocity (Upp & LaNasa, 2002).

Gas flow is often measured using an orifice plate similar to that shown in Figure 7.8b. The pressure drop depends on the size of the orifice, which is selected for a particular application; smaller orifices are generally used for low flow rates. The flow rate is calculated from equation 7.34 using the appropriate proportionality constant (which is provided by the vendor).

Suggested Reading

Asher, A.C. (1997). *Ultrasonic sensors for chemical and process plant.* Bristol, England: Institute of Physics Publishing.

Baker, R.C. (2000). *Flow Measurement Handbook: Industrial Designs, Operating Principles, Performance, and Applications.* New York: Cambridge University Press.

Eggins, B.R. (2002). *Chemical Sensors and Biosensors*. Hoboken, NJ: Wiley.

Fraden, J. (2004). *Handbook of Modern Sensors; Physics, Designs, and Applications,* Third Edition. New York: Springer.

Lipták, B.G. (Ed.) (2003). *Instrument Engineers' Handbook, Volume 1: Process Measurement and Analysis,* Fourth Edition. Boca Raton, FL: CRC Press.

McMillan, G.K. and Considine, D.M. (Eds.) (1999). *Process/Industrial Instruments and Controls Handbook,* Fifth Edition. New York: McGraw-Hill.

8

Particle Size

8.1 Introduction

Particulate matter (in the form of powders, emulsions, aerosols, slurries, etc.) is involved in the majority of commercial manufacturing processes. The particles are discontinuous bits of matter that may be solid, liquid, or gas, and they might be of any size (nanometers to centimeters in diameter). The ability to characterize particles and systems of particles is important because for industrial processes, particle properties directly influence processing (e.g., rheology, mixing, reaction rate) as well as ultimate product performance. These processes are difficult to operate and control because the particle properties can fluctuate significantly throughout the process. Changes in the composition and rate of feed cause process fluctuations, and the resulting impact on size distribution (for example) can lead to production of off-spec material or equipment downtime.

In order to provide high throughput or capacity, many industrial processes operate at high concentrations (e.g. over 20% by weight) that can affect the size distribution and increase the possibility of equipment plugging. Success in controlling such processes requires measurement sensors that can operate at high concentrations. The sensors described in this chapter are designed to measure the particle size distribution (PSD), which is an important characteristic of particulate systems (Scott, 2003). There are literally hundreds of instruments available for measuring PSD, and the reader is referred to Allen (1997) for a more comprehensive introduction to particle size measurement.

8.1.1 Representation of Particle Size

The measurement of particle size is complicated by two issues. First, the concept of "size" for a three-dimensional object is ambiguous and must be precisely defined before any measurement can be made. In the case of spheres (such as droplets in an emulsion) the size usually refers to the diameter, but for most processes the particles are more complicated or even irregular in shape. For instance, certain crystals grow preferentially in one direction, leading to a needlelike shape; depending on the application, the "size" of this crystal might refer to the width, the length, or the volume. Second, the particles in a sample typically come in a range of sizes, however size is defined. Therefore it is necessary to measure the particle size distribution, which is often described by a histogram of particular sizes encountered in a sample. If the PSD has a simple shape, this distribution can be described by a few simple numbers such as the median size (defined below) and the width of the distribution. The layman has a tendency to refer to the median size as the "particle size," and in many cases this information is adequate; however, one should note that the sample may contain particles much larger (or smaller) than the median.

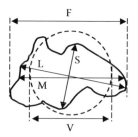

The difficulty of assigning a definite size to an arbitrarily shaped particle is illustrated in Figure 8.1. We could measure the major and minor axis lengths, the aspect ratio (the ratio of these lengths), the projected area, or any number of other quantities. In order to resolve the ambiguity inherent in the concept of particle size we must define whether we are using Feret's diameter or Martin's diameter (defined below),

FIGURE 8.1 Various definitions of size for an irregularly shaped particle: Feret's diameter (F), the minor axis of the projection (S), the major axis of the projection (L), Martin's diameter (M), and the diameter of the equivalent sphere (V). All of these dimensions except V depend on the orientation of the particle.

the diameter of an equivalent sphere, or some other basis. The most commonly used definitions are the following:

Volume diameter: The diameter of a sphere having the same volume as the particle.
Surface diameter: The diameter of a sphere having the same surface area as the particle.
Surface volume diameter: The diameter of a sphere having the same surface-to-volume ratio as the particle.
Stokes's diameter: The diameter of a free-falling sphere traveling at the terminal velocity in a fluid.
Sieve diameter: The width of the minimum screen aperture through which the particle will pass.
Projected area diameter: The diameter of a circle having the same area as the projected area of the particle at rest in a random orientation.
Feret's diameter: The distance between pairs of parallel lines that are tangent to the projected outline of the particle (i.e., the caliper distance).
Martin's diameter: The length of the chord that bisects the projected area of the particle.

In industry, PSD results are often quoted in terms of the volume-weighted diameter because it is equivalent to a mass-weighted diameter, and materials are usually added on a mass basis. Additional information on the definition of particle size can be found in the international standard ISO 9276-1 (1998).

As noted above, samples usually contain particles of various sizes; as an example, consider the hypothetical sample of Figure 8.2. As shown in Figure 8.3, there are several ways

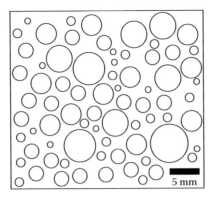

FIGURE 8.2 A hypothetical sample of spherical particles.

to depict the PSD of this sample. The distribution can be shown in terms of a differential plot (sometimes called a *density function*) or a cumulative plot. The differential size distribution of Figure 8.3a, sometimes denoted as $q(x)$, shows the relative probability that a given particle in the sample will have the equivalent diameter x. The cumulative

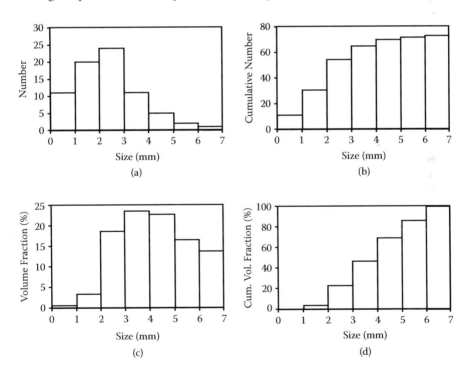

FIGURE 8.3 Particle size distributions for the sample shown in Figure 7.2: (a) number-based differential PSD; (b) number-based cumulative PSD; (c) volume-based differential PSD; (d) volume-based cumulative PSD.

distribution of Figure 8.3b, sometimes denoted as $Q(x)$, shows the relative probability that the particle will have an equivalent diameter equal to or smaller than x. It should be evident that $Q(x)$ can be obtained by summing (or integrating) $q(x)$ over all sizes, and $q(x)$ is obtained by differentiating $Q(x)$ with respect to particle size x. The data depicted in Figure 8.3a and Figure 8.3b count the number of particles, which is a useful representation for applications such as contamination monitoring. Since most industrial operations add raw materials by weight, a more useful representation is usually the volume fraction (which is equivalent to the mass or weight fraction). The volume-based distribution can be calculated by multiplying the number of particles observed at each size x by the equivalent sphere volume and dividing this product by the total volume of particles in the sample. Since the volume is proportional to the cube of the diameter, it should be evident that volume-based distributions tend to appear skewed toward large particle sizes. The volume-based results for $q(x)$ and $Q(x)$ are shown in Figure 8.3c and Figure 8.3d. The shape of the PSD clearly depends upon not only the definition of size but also how it is represented.

Although many instrument manufacturers represent PSD measurements as continuous curves, it should be noted that the data itself is a set of values $\{q_1, q_2, \dots q_N\}$ estimated at a number N of discrete particle sizes $\{x_1, x_2, \dots x_N\}$. The particle sizes $\{x_i\}$ actually represent size classes (sometimes called *bins*) so that all particles larger than x_{i-1} but smaller than or equal to x_i are deemed to have a size of x_i. Alternatively, the size bins may be centered so that the lower bound is $(x_{i-1} + x_i)/2$ and the upper bound is $(x_i + x_{i+1})/2$. In addition, since the PSD often approximates a log-normal distribution (introduced in section 8.5), the sizes are not equally spaced but become progressively farther apart: $(x_i - x_{i-1}) < (x_{i+1} - x_i)$. Due to this irregular spacing, the size axis of a PSD plot is usually logarithmic.

Under ideal circumstances, one or two numbers are sufficient to characterize the PSD. For instance, a gross indication of particle size is given by the mean (or average) diameter, which is calculated from the differential distribution by the weighted average shown in equation 8.1.

$$Mean = \sum_{i=1}^{N} q_i x_i$$

(8.1)

Here it is assumed that the PSD is normalized so that the sum over all $\{q_i\}$ equals one. Depending on the way particles are counted (e.g., number versus volume), the *mean diameter* can refer to the length mean diameter, surface mean diameter, or volume mean diameter. The modal diameter can also be determined from the differential size distribution: it is the most frequently occurring size value and corresponds to the peak of the PSD curve. Sometimes there is more than one peak in the plot of $q(x)$; such distributions are called *bimodal* or *multimodal*.

The cumulative plot $Q(x)$ can be used to determine other characteristic sizes to represent the PSD. A frequently quoted figure is the median diameter, which corresponds to the 50th percentile in the cumulative size distribution. International standard nomenclature denotes it as x_{50}; for historical reasons the median size is often called the d_{50}, but both notations refer to the same quantity. One can get a sense of the width of the

PSD by noting other percentiles in the $Q(x)$. Typically the PSD report includes the x_{10}, x_{50}, and x_{90} (d_{10}, d_{50}, and d_{90}) sizes, and the ratio $(x_{90} - x_{10})/x_{50}$ gives the relative width of the distribution. Other important parameters are x_{16} and x_{84}, which are used to determine the geometric standard deviation σ of the distribution. If the PSD is a log-normal distribution (see section 8.5), then it can be shown that

$$\sigma = \sqrt{\frac{x_{84}}{x_{16}}} \qquad (8.2)$$

8.1.2 Types of PSD Instruments

A wide variety of instruments are available for measuring PSD, and a discussion of each type is quite beyond the scope of this book; interested readers may refer instead to the comprehensive survey by Allen (1997). It is sufficient to note in passing that particle size can be measured using a wide variety of physical phenomena, including light, sound, electricity, X-rays, and fluid mechanics. It is possible to classify these techniques into two groups: those that measure individual particles one at a time, and those that measure an ensemble of particles all at once. Examples of both approaches are given in this chapter. The discussion here will be confined to particles in suspension—that is, ones that are dispersed in a liquid medium.

The size scale of interest to industry ranges from about 1 nanometer up to several centimeters, which encompasses seven orders of magnitude. The solids concentration also varies significantly, from parts per million to well over 50% by volume. No one instrument can cope with such a wide range of material properties, so it is common to choose the instrumental technique best suited for the particular application. Figure 8.4 compares the working range for several techniques that are described in more detail below.

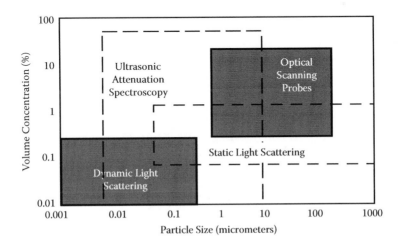

FIGURE 8.4 A comparison of PSD instruments.

8.2 Particle Counting

One approach to measuring the PSD is to determine the size of individual particles one at a time. Since samples generally contain millions of particles, this approach can be time consuming; however it is the only sure way to detect relatively large or small particles that are present in small numbers. The particle size can be detected by using either electrical or optical methods, as described below.

8.2.1 Electrical Counting

Techniques for measuring individual particles in a suspension use the suspension fluid as a vehicle for transporting the particles through or the detector. An early device of this type is the Coulter counter, which was designed to count red and white cells in a sample of blood.[1] In that system, a blood sample is diluted in an electrolyte solution and pumped through a small aperture (Figure 8.5) that separates two electrodes between which an electrical potential is maintained. The aperture allows ions in the electrolyte to conduct current from one electrode to the other. Each time a blood cell passes through the aperture, it partially blocks the ion pathway and momentarily decreases the current by an amount proportional to the cell's cross-sectional area. One measures the particle size by converting the current pulses to voltage pulses, measuring their amplitude, and sorting the results according to pulse height (using a multichannel analyzer or equivalent instrumentation). Assuming the system has been calibrated against a particle size standard, the resulting number-based distribution is the PSD.

Since the basic Coulter counter requires that the sample be diluted in a weak electrolyte, it is not suitable for industrial applications where the particle dispersion (hence the effective particle size) might be affected by the electrolyte.

8.2.2 Optical Counting

A similar technique is based on the blockage of light rather than electrical current to measure the particle size; this technique can measure particle size in virtually any

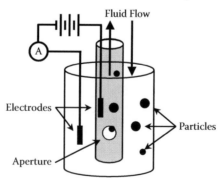

FIGURE 8.5 A diagram of a Coulter counter. When particles in the electrolyte are pumped through a small aperture in the tube, they affect the current passing between the two electrodes.

transparent liquid. Optical particle counters are similar in many respects to a Coulter counter, except that they measure light obscuration rather than fluctuations in current. A very dilute sample, on the order of 100 parts per million (ppm), passes through a capillary tube that is illuminated by a solid-state laser. Shadows cast by the particles fall on a detector, and the concentration is sufficiently low to ensure that only one particle at a time is casting a shadow. The size of the particle determines the size of the shadow; therefore the electrical pulse generated by the detector reflects the size of the particle. Over time, the pulses are measured and sorted according to height by a multichannel analyzer, which generates a histogram showing the PSD. Optical counters typically operate in the range of about 1–400 μm.

Due to the discrete and random nature of the measurement process, the count in each size bin is subject to Poisson counting statistics. Therefore, the error associated with a particular count of N is on the order of $\sqrt{N}$. A count of 100 is really 100 ± 10 (10% error), whereas a count of 10,000 has an uncertainty of 100 (only 1% error). For this reason, it is necessary to ensure that a sufficiently high total count is achieved so that the measurement is not marred by the counting statistics.

The maximum count rate of the system is limited by both the response time (bandwidth) of the detector and the physical spacing between particles in the suspension. If two distinct particles are too close, the pulses pile up in the preamplifier, which means that the voltage level does not have enough time to return to the baseline before the next pulse arrives. The two pulses appear as a single misshapen pulse (as shown in Figure 8.6), and the system counts them as a single large particle. This situation is known as *coincidence* and is a well-known problem in the measurement of radioactivity and other stochastic phenomena. For relatively low amounts of coincidence, the PSD will have a few counts at very large particle sizes. Frequent coincidence events will cause the entire measured PSD to broaden and shift toward larger sizes, so it is important to measure only dilute samples. Under normal operating conditions, the pulses in the detector output signal are well-separated, as depicted in Figure 8.7. As a rule of thumb, these instruments measure particles at the parts per million level.

Several issues must be considered in order to make meaningful measurements with optical counters. First, the dispersion, dilution, and injection of the sample must be perfect in order to calculate the counts per milliliter (mL) correctly; the absolute count data must therefore be treated with some skepticism. Second, the flow rate through the measurement cell has been observed to fluctuate by over 10% in certain commercial

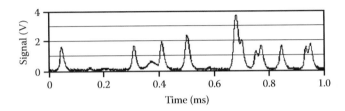

FIGURE 8.6 Pulse output from the optical detector when the particle concentration is too high; several double peaks are evident.

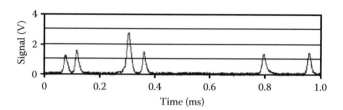

FIGURE 8.7 Pulse train observed when the sample used to obtain Figure 8.6 is diluted to normal operating conditions.

particle counters, so the total volume of suspension measured by the instrument may be incorrect. Finally, the detection efficiency is a function of particle size, especially for the smaller particles. When combined with the inherent uncertainty in counting random events, these issues make it difficult to obtain reliable results. Experience has shown that one way to extract meaningful information is to normalize the counting data, as explained in the next section.

8.2.3   The Normalization of Counting Data

Normalizing the data can reduce measurement-to-measurement variability caused by the issues listed above. This procedure seeks to improve the statistics by combining size bins in the area of interest, and it also reduces the error due to variability in sample concentration and flow rate. It is especially useful for interpreting the coarse tail of the size distribution, even when the median size is too small to be measured accurately with an optical counter.

Consider the raw data set $\{q_1, q_2, \ldots, q_N\}$ generated by counting the number of particles in N distinct size classes; this data is the discrete representation of the number-based differential PSD, $q(x)$. In the coarse tail of the PSD, these numbers will be quite small (typically less than a few hundred counts in each size bin) so the counting statistics are poor. Many applications require an assessment of the amount of particles larger than a given size, and in those cases the statistics can be improved by pooling the "oversize" counts.

Since the data is discrete we can sum (rather than integrate) over the particle size to obtain a cumulative count C_k of particles that are equal to or larger than particle size class k:

$$C_k = \sum_{i=k}^{N} q_i$$

(8.3)

The definition for C_k is similar to $Q(x)$, the number-based cumulative PSD, except that it counts the particles larger than a given size rather than those that are smaller. As an example, Figure 8.8a shows the cumulative oversize data from replicate measurements of a titanium dioxide (TiO_2) powder sample. The median particle size is about 0.3 μm, so only the coarse tail can be seen. The sample was prepared at a concentration of 0.6 mg/mL in an aqueous solution of 0.1% by weight (wt) tetrasodium pyrophosphate, and it was stirred continuously to prevent sedimentation. Both measurements shown in

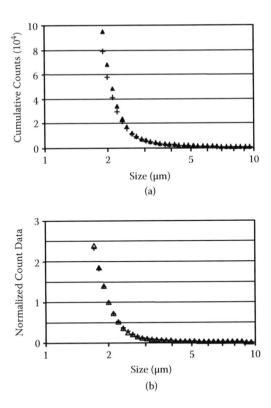

FIGURE 8.8 Optical counting results for replicate measurements of oversized TiO2 particles: (a) the cumulative oversize data shows a significant discrepancy between the measurements; (b) the same data after normalization according to equation 8.4.

Figure 8.8a were made by taking a 100 μL aliquot of this prepared sample and injecting it into the optical counter, which mixed the aliquot into an additional 35 mL of ultrafiltered water. The solids concentration in the counter was therefore 1.7 μg/mL. Although both measurements were performed with great care, there is an obvious discrepancy between the raw data sets in the 1–3 μm region. One measurement yields 68,000 particles larger than 2 μm, and the other yields 58,000 particles larger than 2 μm.

The data depicted in Figure 8.8a can be normalized by choosing a basis size x_m to provide a reference point and defining the particle fraction $R_{k,m}$ as the ratio of the cumulative count to the cumulative count at the basis size:

$$R_{k,m} = \frac{C_k}{C_m} = \frac{\sum\limits_{i=k}^{N} q_i}{\sum\limits_{i=m}^{N} q_i} \tag{8.4}$$

A plot of $R_{k,m}$ over all values of k gives the normalized cumulative oversize fraction. Figure 8.8b shows the results of applying equation 8.4 (with a basis size of 2 μm) to the data plotted in Figure 8.8a, and it can be seen that this normalization considerably improves the agreement between the two data sets.

Another example of data normalization comes from the analysis of gold powder used to make conductive paste for electronic applications. Nine different lots of gold powder were analyzed, with two identical optical particle counters located in separate laboratories. The number of oversized particles (calculated with equation 8.3) in each was divided by the total number of counted particles to determine the fraction larger than 4, 5, and 6 micrometers. The results from both laboratories are compared in Figure 8.9a, where the results from one lab are plotted versus the corresponding results from the other. The fractions larger than 4 μm, 5 μm, and 6 μm are depicted by the squares, triangles, and diamonds, respectively. There is clearly a considerable discrepancy between the raw data sets provided by the two labs: a linear regression reveals that the R^2 correlation factor is low (less than 80%), and the slope of the line is 0.42 (it should have been 1.00). After normalization (the results of which are shown in Figure 8.9b), the data sets showed excellent agreement. The correlation factor between the normalized data sets is 98%, and the slope of the linear regression is 0.996, so the systematic error between the labs is only 0.4%.

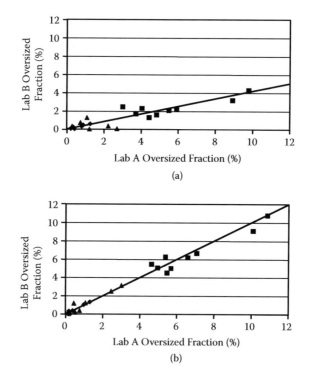

FIGURE 8.9 Data normalization: (a) direct comparison of oversized fractions in gold powder, measured with two identical particle counters located in separate laboratories; (b) comparison of the same data sets after normalization according to equation 8.4.

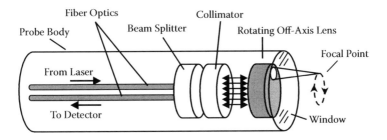

FIGURE 8.10 Operation of the FBRM probe. The off-axis rotating lens scans the laser beam in a circular pattern, and the back-scattered light is focused onto a fiber optic that carries it to a photodetector.

Thus, particle counting methods can provide reproducible results, provided sufficient care is given to both the measurement and the interpretation of the results.

8.2.4 Focused Back-Reflection Method

The particle counters described above use the flow of the suspension fluid to carry the particles past the detector. A different optical instrument, capable of measuring particle sizes at higher concentration, is based on a laser scanner that sweeps a focused beam of light past the particles. The Lasentec FBRM (focused back-reflection method) optical probe can be used to monitor the PSD in the range 1–1000 μm (Preikschat & Preikschat, 1989).[2]

The operation of the FBRM probe is illustrated in Figure 8.10. Light from a laser diode is focused through a sapphire window into the process where it forms a spot (about 0.8 μm in diameter). Rotating optics in the probe scan this spot in a circular path at a fixed tangential velocity (about 2 meters per second [m/s]); when this scanning spot strikes a particle, the light is reflected back into the probe and is detected with a photodiode. The resulting back-scattered signal (depicted in Figure 8.11) is analyzed to determine

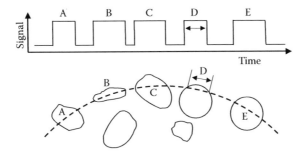

FIGURE 8.11 Chord-length measurement with the FBRM probe. The circular scan path of the focused laser beam is depicted with a dashed line. As the focal spot crosses over each particle, the back-scattered light is detected and the output signal of the photodetector momentarily goes to a high level. The chord length is proportional to the pulse width of the detected signal. Note that particles D and E are the same size, yet give different chord lengths. Particles not intersected by the scan line are not seen by the probe.

the width (*chord length*) of the pulses, which are proportional to the size of the particles encountered by the scanning spot.

Due to motion of the particles, no two scans yield the same chord length, so the instrument generates a distribution of chord lengths even for a monodisperse system. To a first approximation, the average chord length is related to Martin's diameter. If the particle shape is known, the chord length distribution can in theory be deconvolved to give a number- or volume-based size distribution. To some extent, changes in particle shape can be monitored by examining the second moment of the distribution. The correct interpretation of the data generated with this instrument must be interpreted with care, and every new application requires thorough study.

The advantage of this probe is that it can measure concentrated suspensions directly in process, provided the flow velocity across the probe tip is much less than 2 m/s (i.e., the scanning speed). A disadvantage is that although the probe is sensitive to changes in the particle size, the results are affected by changes in the process flow velocity, particle morphology, and solids concentration.

8.3   Optical Scattering Techniques

8.3.1   Static Light Scattering

The *static light scattering* (SLS) technique, sometimes called *laser diffraction* or *low-angle scattering*, is arguably the most popular method for measuring particle size in both wet and dry applications. SLS instruments are widely used in industrial laboratories, and they are available from a large number of vendors. They generally can measure particle size from less than 100 nm to over 1 mm. The measurements are accurate, and the technique is well understood (see international standard ISO 13320-1: 1999); the main drawback is that opaque or highly concentrated suspensions cannot be measured. In online applications, this limitation can sometimes be met by diluting a side stream with additional carrier fluid.

The basic concept of laser diffraction is shown in Figure 8.12, which depicts a collimated laser beam illuminating a single particle in a moderately dilute suspension that flows through a transparent cell. The particle diffracts or otherwise scatters the light to form a pattern of intensity $I(\theta)$, where θ denotes the scattering angle measured with respect to the direction of the incident light. It is known from the theory of light scattering (Mie, 1908) that the size of an individual particle determines the observed $I(\theta)$ pattern, and an ensemble of particles will create a composite pattern from which the original size distribution can be inferred.[3] At every moment the instantaneous diffraction pattern is therefore the composite response from the thousands of individual particles illuminated by the beam. A lens collects this light and focuses it on an array of photodetectors (mounted in the focal plane) in such a way that all the light scattered at a particular angle θ is focused in a ring of a specific radius r (see the discussion on lenses in chapter 4).[4] Thus, the lens enables the entire ensemble of illuminated particles to be analyzed, and the diffraction pattern appears as a series of concentric rings.

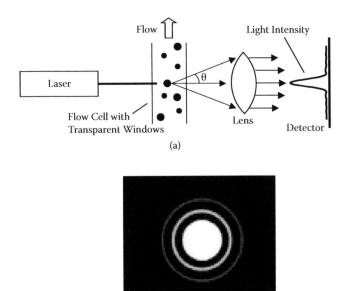

(a)

(b)

FIGURE 8.12 The static light scattering technique. The laser light scattered by particles (a) forms a diffraction pattern (b).

The detector array records the radial variation in the intensity of the light, $I(r)$, where for a lens of focal length f,

$$r = f \tan\theta \qquad (8.5)$$

In practice, the diffraction pattern is integrated over a period of several seconds (or equivalently, averaged over many measurements) in order to improve the signal-to-noise ratio. If fresh sample flows through the cell, this integration also ensures a representative measurement by expanding the sample size to include many more particles. The diffraction pattern is then analyzed by using an optical scattering model based on either the Mie theory or the Fraunhofer approximation.

Mie scattering theory is a comprehensive and completely general description of light scattering by spherical particles. It is a first-principles theory based on two assumptions: the particles scattering the light are assumed to be spheres, and the possibility of multiple scattering is neglected. This second assumption means that the results are valid for the single-scattering regime only and cannot reliably describe scattering in concentrated suspensions. The theory can be extended to consider scattering from particles with different shapes and aspect ratios. The calculation requires the real and imaginary components of the refractive index of the particle and the real component of the refractive index of the fluid.[5]

The Fraunhofer approximation is applicable when the diameter of the particle scattering the incident light is much larger than the wavelength λ of the radiation. The approximation can generally be used to measure particles larger than about 5 μm, and it conveniently does not require the refractive index of the particle. This feature is particularly useful for analyzing mixtures of particle types, where the index of refraction could vary from one particle to the next.

The Fraunhofer approximation can be derived as a limiting case of the Mie scattering theory, or it can be derived independently by considering the diffraction of light from two opposite edges of the particle. The phase difference between the two diffracted beams results in an interference pattern marked by a series of maxima and minima. When all of the azimuthal angles are considered these interference patterns combine to form a ringlike pattern of light scattered in the forward direction. The angular dependence of the resulting diffraction intensity pattern is

$$I(\theta) \propto \left[\frac{J_1(b\sin\theta)}{b\sin\theta} \right] \tag{8.6}$$

where the constant b is defined by

$$b \equiv \frac{\pi x}{\lambda} \tag{8.7}$$

and x is the particle diameter; J_1 is the Bessel function of the first order. The first zero of the intensity function occurs at the angle θ_0 given by

$$\theta_0 = \sin^{-1}\left(\frac{1.22\lambda}{x} \right) \tag{8.8}$$

Therefore, the spacing between the rings can be used to determine the particle size.

One might imagine that these optical models could be used to extract the PSD directly from the observation of $I(r)$; however, direct inversion does not generally work well with real data sets, as discussed in section 8.5. Instead, the PSD estimation algorithm generally uses a scattering model to calculate the pattern $I(r)$ based on an assumed PSD, $q(x)$. The estimate of $q(x)$ is adjusted iteratively until the calculated $I(r)$ equals the measured value to within a predefined error tolerance. At that point the estimated $q(r)$ is taken to be the PSD of the sample. Most commercial instruments use this type of approach.

Since laser diffraction measures an ensemble of particles, it tends to ignore particles that are significantly larger or smaller than the rest, especially when the outliers occur in small quantities. The scattering pattern they generate is simply too weak to be identified. An example of this deficiency can be seen in Figure 8.13, which shows the PSD obtained for an experimental grade of TiO_2. The sample is known to have a coarse tail extending to well over 10 μm as determined by particle counting and optical microscopy, but the largest particle detected via static light scattering is only 3.55 μm. This result is included as a caution to those who are using laser diffraction techniques in applications where the top size is critical.

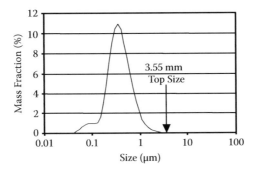

FIGURE 8.13 Static light scattering may not detect the smallest or largest particles in a sample.

8.3.2 Dynamic Light Scattering

Colloids and nanoparticles, which are well under 100 nm in size, are difficult to measure with the SLS technique for the simple reason that they are so small that they do not scatter very much light. Lord Rayleigh (John Strutt) showed that the scattering cross-section for particles that are much smaller than the wavelength of light is proportional to the sixth power of the particle diameter (Strutt, 1871). Therefore, a 50 nm particle scatters only about 1.5% of the amount of light scattered by a 100 nm particle. As the amount of scattered light decreases, the contrast in the pattern of bright and dark rings created through scattering also decreases until the pattern can no longer be detected reliably.

Dynamic light scattering (DLS), which is also called quasi-elastic light scattering and photon correlation spectroscopy, examines the frequency of scattered light rather than variations of light intensity in order to measure particle size.[6] The technique is a practical application of Brownian motion, which is the random and microscopic motion seen in a dilute suspension of fine particles roughly 1 μm in size or smaller. This phenomenon was thoroughly described by the botanist Robert Brown in 1828, but the mechanism behind it was not understood until Albert Einstein's 1905 paper on the motion of small particles suspended in a liquid. The motion is due to thermal energy, which causes particles to diffuse through the suspension. In the case of spherical particles, Einstein found the diffusion coefficient D to be

$$D = \frac{kT}{3\pi\eta x} \tag{8.9}$$

where k is Boltzmann's constant, T is temperature (Kelvin), η is the viscosity, and x is the diameter of the particle. Thus one can estimate the particle size by measuring the diffusion coefficient and solving equation 8.9 above. Note that the viscosity of the fluid must be known, and since this parameter is sensitive to temperature, it is important to measure the particle diffusion at a stable temperature for which the viscosity is known.

The diffusion coefficient D has units of $[m^2 \ s^{-1}]$ and is defined as the proportionality constant in Fick's Law that relates the local concentration's time rate of change to its second order partial derivative with respect to position. If the particles are small, the value of

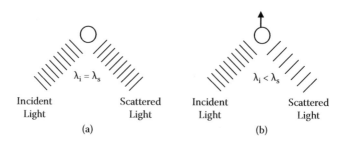

| | (a) | | (b) |

FIGURE 8.14 Light scattering from a particle: (a) the wavelength of scattered light does not change when the particle is at rest; (b) the Brownian motion of submicron particles introduces a Doppler shift in the scattered light. Here the particle is moving away from the source and detector, so the wavelength increases.

D predicted by equation 8.9 will be relatively large, and the particles will diffuse quickly through the liquid. At any instant their Brownian velocities, though oriented in random directions, will be relatively high (on the order of 1 cm/s).

The velocity due to the Brownian motion can be measured by reflecting coherent light from the particles and examining the Doppler shift in the scattered light, as illustrated in Figure 8.14. The Doppler effect (see sections 3.4.5 and 4.2.7) causes the frequency of the light to be shifted by an amount that is proportional to the speed of the particle. The relative change in frequency is rather small: visible light has a frequency of about 0.5 THz $(5 \times 10^{14}$ Hz), whereas submicron particles at room temperature change the frequency of the light by only 1–10 kHz. Thus, the relative Doppler shift is roughly 1 part in 10^{12}, and an extremely precise measurement technique is needed to detect it.

The optical heterodyne shown in Figure 8.15 provides a means of detecting the frequency shift via interferometry. A laser beam is split into two: one beam is scattered by particles in the suspension and recombined with the other beam on an optical detector, which is typically a photomultiplier. The interference between the two beams causes the intensity at the detector to pulsate at the beat frequency, which is simply the difference in frequency $(f_1 - f_2)$ between the original and scattered beams. Thus, the beat frequency

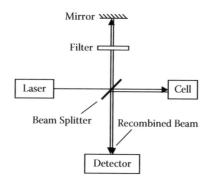

FIGURE 8.15 A schematic of an interferometer for measuring dynamic light scattering.

is equal to the Doppler shift caused by Brownian motion of the particles. Of course, this beat signal $s(t)$ continually changes with time because there are many particles, and each one moves randomly. Defining the autocorrelation function $G(\tau)$ of $s(t)$ to be

$$G(\tau) = \int_{-\infty}^{\infty} s(t)s(t-\tau)dt \qquad (8.10)$$

it is found that $G(\tau)$ describes an exponential decay function

$$G(\tau) = A + Be^{-2\Gamma\tau} \qquad (8.11)$$

where the decay constant Γ is given by

$$\Gamma = Dk^2 \qquad (8.12)$$

and

$$k = \frac{4\pi n}{\lambda_0}\sin\left(\frac{\theta}{2}\right) \qquad (8.13)$$

Here, θ is the observation angle, λ_0 is the wavelength of the incident light, n is the refractive index of the fluid, and A and B are constants. Therefore, to apply this technique one measures $s(t)$, computes $G(\tau)$ using equation 8.10, fits the exponential decay of equation 8.11 to the graph of $G(t)$ versus τ to find Γ, and finally determines D from equations 8.12 and 8.13. The particle size is then determined from equation 8.9, assuming the particle is spherical. For polydisperse systems, $G(\tau)$ is a superposition of decay curves, each with its own decay constant.

As mentioned at the beginning of this section, optical techniques cannot be used directly in applications that involve concentrated or opaque suspensions. Likewise, they cannot be used for in-process applications where the sample might be expected to coat or foul the windows of the sample cell. In order to address those situations, a different sensing mechanism is needed; the next section introduces the use of ultrasound as an alternative technique that can be used to measure PSD in a wide variety of industrial processes.

8.4 Ultrasonic Attenuation Spectroscopy

It has long been understood that ultrasound can be scattered by particles, and that the scattering is affected by the particle size. Rayleigh, in his 1877 *Theory of Sound*, derived an expression for the scattering intensity in the long wavelength limit and predicted that it is proportional to the square of the particle's volume (Strutt, 1945, p. 277). More recently, a significant effort has been underway to adapt ultrasonic spectroscopy to the measurement of particle size distributions in slurries and suspensions.

The basic concept, illustrated in Figure 8.16, is to measure either the frequency-dependent attenuation or the phase velocity of ultrasound as it passes through the sample. The ultrasound is generated by applying a high frequency electrical signal to a

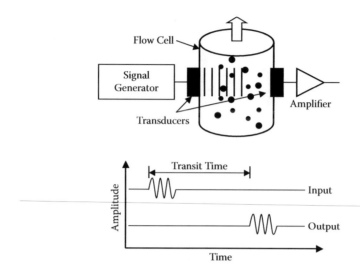

FIGURE 8.16 A diagram of an ultrasonic measurement, showing the transit time of a wave moving through a suspension of particles.

piezoelectric transducer, which constricts in response to the electrical signal and generates an ultrasonic wave. The wave propagates through the suspension and is received by another transducer, which converts the energy back into an electrical signal for analysis. By varying the frequency of the initial ultrasonic wave, one obtains a spectrum of attenuation or phase velocity (or both) as a function of frequency. This data is related to the PSD and concentration of the sample by empirical observation or by theoretical calculations. Once this connection is established, one can in principle estimate the PSD from the ultrasonic data. The focus here will be on ultrasonic attenuation spectroscopy rather than on phase velocity measurements.

Although the theoretical basis of ultrasonic scattering has been known for many years (Riebel & Löffler, 1989; Pendse & Sharma, 1993), most people (even those working in the field of particle characterization) are not familiar with the use of ultrasound to measure particle size. Therefore, the industrial application of ultrasonic spectroscopy is discussed below in some detail; the reader may also refer to the suggested reading at the end of this chapter.

8.4.1 Theoretical Background

As ultrasound propagates through a suspension of particles, the amplitude and phase of the component frequencies are affected by material properties of the system, including the particle size distribution. A theoretical framework is needed to interpret the observed changes in the ultrasonic signal. In the present discussion, predicting the frequency-dependent attenuation caused by a suspension, emulsion, or slurry with a known PSD is called the *forward problem*. The *inverse problem* of estimating the PSD from the observed ultrasonic attenuation spectra is discussed in section 8.5.

The ultrasonic technique described below is based on theoretical models that have been developed over the past century, starting with the work by Sewell (1910) on sound absorption in fogs. Urick (1948) showed that ultrasonic absorption in aqueous suspensions is largely due to viscous drag between the fluid and the particles, and Epstein and Carhart (1953) included the effects of particle motion and thermal conduction in aerosols. Allegra (1970) demonstrated that Epstein and Carhart's theory was not limited to aerosols, and his work with Hawley (1972) has provided a mathematical framework for calculating the attenuation of ultrasound in suspensions. The combined theory of frequency-dependent attenuation due to scattering of ultrasound is now known as *ECAH theory* (named for Epstein, Carhart, Allegra, and Hawley), the development of which has been reviewed in detail elsewhere (Challis et al., 2005).

The primary benefit of ultrasound is that it can pass through highly concentrated particulate systems, such as slurries. Holmes and Challis (1993), and Holmes, Challis, and Wedlock (1993) have measured absorption and phase velocity in monodisperse polystyrene suspensions of up to a volume fraction of 45% and found good agreement with theoretical predictions. Atkinson and Kytömaa (1993) have investigated two-phase mixtures of large (1 mm) silica beads in water and showed that the attenuation has a maximum value near 30% by volume. As described later in this chapter, the author has measured the particle size in pigment slurries up to volume fractions of about 45%, over an acoustic path length of 5 cm (Scott et al., 1998). The significance of being able to measure PSD over such a distance in a slurry is that it validates the practicality of this approach for use in an industrial process.

The attenuation of ultrasound tends to increase with frequency, so that at some point the ultrasonic signal becomes too small to measure reliably. For this reason, practical sensors for industrial process applications tend to operate in the frequency range where the ultrasonic wavelength is much longer than the particle diameter x. This regime is dominated by viscous effects, and the attenuation is expected to increase as the square of the frequency.

Attenuation in the long wavelength regime is adequately described by the ECAH model, which considers a plane pressure wave impinging on a spherical particle (Figure 8.17). The compression, shear, and thermal waves produced in the particle can be described by three wave equations, each with its own complex wave number. The waves are then expanded in partial waves (spherical harmonics), with separate wave functions in the fluid and in the particle. By applying the boundary conditions at the surface of the sphere and invoking the orthogonality of the basis functions used in the expansion, one arrives at a set of six coupled equations in six unknowns for each term in the partial wave expansion.

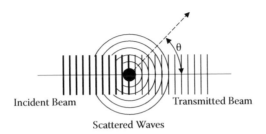

FIGURE 8.17 The scattering of a plane wave of ultrasound impinging on a sphere.

Fortunately, the series converges quickly and in many cases only the first two terms are needed to describe the attenuation. The equations in the original paper contained errors that have since been corrected by Challis et al. (1998, 2005). Povey (1997, appendix B) has published a MathCad spreadsheet that calculates the explicit solutions of the ECAH theory in the long wavelength limit.

The ECAH model uses a total of 16 physical constants, many of which are temperature-dependent. It is often difficult to find reliable values of all 16 parameters, especially in the case of compounds or new materials. Fortuitously, in the case of many industrially significant materials only a few of these constants need to be known accurately. Table 8.1 shows the results of a variational calculation of the attenuation in a suspension of a relatively dense pigment (TiO_2) in water. Here the relative sensitivity of the model to each of these parameters at 25°C is shown as a percentage change in the calculated attenuation due to a 1% increase in the value of that parameter (holding the others fixed). The resulting change in attenuation is a function of frequency, so the values given in Table 8.1 are approximate values (within an order of magnitude) over the 10–100 MHz frequency range. In this case (a rigid, dense particle that is smaller than the wavelength) it is seen that the most critical parameters are the sound speed, density, and viscosity of the

TABLE 8.1 Approximate Sensitivity of the ECAH Model to Errors in the Physical Constants in the 10–100 MHz Frequency Range (at 25°C) for a Suspension of TiO_2 in Water

Properties of Fluid	Sensitivity (%)
sound speed	+1.03
density	-2.25
shear viscosity	+0.44
thermal conductivity	+0.0084
heat capacity	-0.016
sound attenuation	-0.002
thermal dilatability	-0.0094
ratio of specific heat	+0.03
Properties of Particle	**Sensitivity (%)**
sound speed	+0.03
density	+1.80
shear rigidity	-0.011
thermal conductivity	+0.003
heat capacity	+0.004
sound attenuation	+0.00005
thermal dilatability	+0.01
ratio of specific heat	-0.01

Note: The relative sensitivity is defined to be the percentage change in calculated attenuation due to a 1% increase in the value of that parameter (holding the others fixed).

suspension fluid and the density of the particle. One can use approximate values for the other physical parameters without sacrificing much accuracy in the results.

The equations given by the ECAH model describe a monodisperse, low-concentration suspension—that is, the particles are all the same size, and they are well separated from each other.[7] The theory predicts the excess attenuation α (measured as decibel [dB] loss per unit length) as a function of ultrasonic frequency f and particle diameter size x. Ultrasound is absorbed in liquids and therefore is attenuated even in the absence of particles to scatter its energy; therefore the term *excess attenuation* refers to the additional attenuation caused by the dispersed phase. In the case of dilute suspensions, the attenuation is proportional to volume concentration c:

$$\alpha(f,x,c)=c\cdot ECAH(f,x) \tag{8.14}$$

Unfortunately, the suspensions typically found in many industrial processes are concentrated and contain particles with a range of particle sizes; this complication is addressed below.

Examples of attenuation spectra predicted by the ECAH model are shown in Figure 8.18 for several sizes of TiO_2 particles in water. This calculation assumes a solids concentration of only 2% by weight, which is roughly 0.5% by volume. The marked variation in attenuation due to particle size suggests that particle size information can be extracted from ultrasonic attenuation spectra.

Polydisperse suspensions are those that contain particles of more than one size. In the case of dilute polydisperse suspensions, each size class contributes to the overall attenuation independently of the others. Thus, the ECAH model can be used to predict the ultrasonic attenuation in a polydisperse suspension of a given concentration by integrating the monodisperse result over the PSD:

$$\alpha(f)=c\cdot \int ECAH(f,x)\cdot p(x)\cdot dx \tag{8.15}$$

where $p(x)$ is simply a function that describes the PSD.

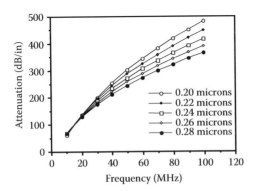

FIGURE 8.18 Calculated attenuation spectra for 2% (vol.) suspension of monodisperse TiO_2.

In the case of high concentrations, physical models based on single-scattering (such as the ECAH model) break down. The effect on attenuation caused by increasing the slurry concentration is shown in Figure 8.19, which depicts the measured attenuation for one grade of TiO_2 slurry for volume concentrations ranging from 0.5% to 38% (or 2% to 72% by weight; see Scott et al., 1995). The excess attenuation data are shown here for 10 MHz (triangles), 20 MHz (circles) and 40 MHz (squares). Apparently there are three distinct concentration regimes. For concentrations below 4 vol% (15 wt%), the attenuation at each frequency is proportional to the concentration of the slurry. The linear response in this single-scattering regime shows that each particle affects the ultrasonic field independently of the others. Dashed lines have been added to the figure to extrapolate this linear response to higher concentrations. In the intermediate regime of 4–10 vol% (15–32 wt%), the observed attenuation is somewhat higher than predicted by a linear theory such as the ECAH model. This result is due to multiple scattering, where the scattered ultrasound itself is scattered before reaching the receiving transducer. Extensions of the linear model have been proposed by various authors to take multiple scattering into account (Lloyd & Berry, 1967; Waterman & Truell, 1961). A third regime, at concentrations over about 10 vol%, is found to have attenuation significantly lower than predicted by a linear model. There is currently no universally accepted, generally applicable theory for predicting ultrasonic attenuation when the interparticle spacing is much less than the particle diameter. Riebel (1992) has found that in the short wavelength limit, a modified Beer-Lambert law can be used to describe the spectrum even at high concentrations, but due to practical limitations this theory cannot be used for submicron particles (as it would require frequencies over 1 GHz). Thus, in the case of concentrated slurries of submicron particles, semiempirical corrections must be applied to ECAH theory to determine PSD from the observed ultrasonic spectra (Hipp et al., 2002a, 2002b; Scott et al., 1998).

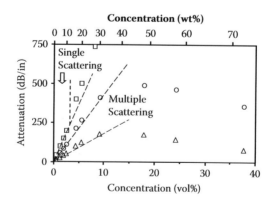

FIGURE 8.19 Attenuation in TiO2 slurries as a function of concentration at 10 MHz (triangles), 20 MHz (circles) and 40 MHz (squares). Adapted from Scott et al., 1995; used with permission.

8.4.2 Hardware Considerations

All theoretical considerations aside, one must measure the ultrasonic attenuation spectrum before it can be used to determine PSD. For industrial applications, especially online measurements, ultrasonic instruments face a number of difficult challenges. These difficulties include:

- the presence of gas bubbles in the process stream
- incomplete filling of the process line
- potential plugging of the flow cell, process line, or side stream
- harsh environment (heat, pressure, vibration, abrasion, corrosion)
- extreme attenuation of the ultrasonic signal

At these frequencies, ultrasound does not propagate very far through gas; therefore, the presence of air bubbles (or a partially filled process line) prevents the ultrasonic signal from being detected. The potential for process material to plug up the instrument flow cell, process line, or side stream is very high. (In fact, this problem provided the motivation to design the instrument described in the next section.) The harsh environment found in industrial plants can cause failure of the ultrasonic transducers and the associated electronic systems; in addition, temperature excursions affect the transducers as well as the interaction between the ultrasound and the process stream. Finally, some process streams (such as highly concentrated slurries of dense submicron particles) so effectively attenuate the ultrasound that measurements on the received signal become difficult.

In view of the obstacles just listed, the design of a robust in-line ultrasonic instrument for measuring PSD in industrial applications should incorporate the following guidelines (Scott, 1998):

- The flow cell must have a large bore to prevent plugging, and the gap between transducers should be as large as feasible.
- Flow through the instrument must be streamlined so that particle segregation does not occur.
- There should few (and preferably zero) moving parts; sliding seals must be avoided when abrasive materials are to be measured.
- Materials of construction must be suitable to the application, as many industrial process streams are highly corrosive.
- In the case of crystallization applications, the flow cell must not act as a heat sink; otherwise, the instrument will become encrusted.
- The placement of the flow cell in the process stream must take into account such factors as complete filling of the pipe, possible entrainment of gas bubbles in the process stream at that point, and the expected temperature and pressure to be encountered.
- In-line placement is preferred over side-streams in most applications for two reasons: adequate sampling is assured, and the potential for plugging is reduced (although the consequences of plugging are more severe).
- The data inversion algorithm must be sufficiently robust to extract PSD from noisy or limited ultrasonic data. The software should produce a goodness-of-fit

parameter that indicates how well the estimated PSD accounts for the observed spectrum; ideally, the PSD estimates should include an estimated error in the quoted values to indicate the reliability of the measurement.
- The instrument should require no operator intervention and minimal calibration.
- The simplicity of the overall design should lead to ease of maintenance.

These principles have been incorporated into the design of an in-line ultrasonic instrument for measuring PSD in concentrated slurries, as described below.

Another consideration is the ultrasonic technique used to measure the ultrasonic attenuation. Most instruments use the through transmission technique illustrated in Figure 8.16, but it is also possible to use the pulse-echo technique, wherein an ultrasonic pulse is bounced from a reflector (or the far wall of a process pipe) back into the transmitting transducer, which also acts as a receiver. In either technique the ultrasound propagates through the suspension for a known distance (which defines the acoustical path length) before it is converted into an electrical signal and analyzed.

A final consideration is the type of electrical signal chosen to excite the transducer. As mentioned in connection with Figure 8.16, the frequency of the ultrasound must be varied in order to obtain the attenuation spectrum. Figure 8.20 depicts three types of ultrasonic signal: swept frequency, tone-burst, and pulsed. Obviously, the desired frequency range can be spanned by sweeping the frequency either continuously (as shown in Figure 8.20a) or by discrete steps. The problem with such an approach is that it sets up a traveling wave, with the echo formed at the far wall bouncing back and forth within the sample cell. The strength of the received signal therefore peaks at the resonant frequencies determined by the acoustic path length and the speed of sound in the suspension. Thus, the interpretation of the received signal is affected by both the design of the cell and the material flowing through it.

A better excitation method, and one that virtually every commercial instrument uses, is to apply a series of short bursts (typically of five to ten cycles) of a sinusoidal voltage, as shown in Figure 8.20b. Consequently, the transducer produces a tone burst with a center frequency equal to that of the sinusoidal wave. The resonances caused by the formation of traveling waves are avoided by keeping the duration of the burst shorter than the

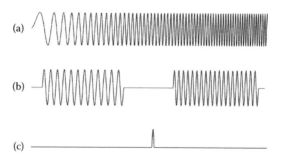

FIGURE 8.20 Types of ultrasonic signal: (a) a swept-frequency signal; (b) a pair of tone-bursts where the frequency is higher in the second tone burst; (c) a broadband pulse containing a wide range of frequency components.

transit time in the suspension and the time between bursts long enough for the echoes to fade completely. (Note that this second requirement extends the time necessary to make a measurement.) Since the excitation is at a single well-defined frequency during the tone burst, it is possible to use a lock-in amplifier or other phase sensitive detection method at the receiver, with the result that the achievable signal-to-noise ratio is superior to that of other methods. By changing the frequency of subsequent tone bursts (note the difference in wavelength between the two tone bursts shown in Figure 8.20b), one builds up a complete spectrum. The tone burst approach does have two distinct disadvantages: the electronic package is complex, and the data acquisition time is often rather long, typically lasting from several seconds up to several minutes. In the case of industrial processes, where the linear flow velocity in a pipe can be faster than several meters per second, it is clear that the sample in the flow cell might change significantly over the course of a single measurement. Put another way, there is no guarantee that the data measured at low frequencies can be combined with the data measured at high frequencies to construct a spectrum that is representative of the particles in the process stream.

For dynamic chemical processes, the only way to ensure that the attenuation spectrum has physical significance is to complete the measurement on a time scale that is short compared to the residence time of the suspension within the flow cell. If such speed can be achieved, then information about the homogeneity or stability of the process stream can be obtained by taking multiple readings in rapid succession. It is possible to record the entire attenuation spectrum in 10–30 μs (depending on the sound speed and acoustic path length) by using a single excitation pulse of very short duration, as shown in Figure 8.20c. This pulse (typically 10–100 ns long) can drive a broadband transducer to produce an ultrasonic signal that contains a wide-band frequency range. It is known from Fourier analysis that the range of component frequencies in a pulse is inversely proportional to the width of the pulse, so a very short ultrasonic pulse interrogates a wide range of frequencies at once. The important distinction between pulse excitation and the tone burst method (with which it is sometimes confused in the literature) is that a single pulse can measure the spectrum, whereas dozens of tone bursts are needed to obtain the same information. Multiple pulses may be used to improve the signal to noise ratio through signal averaging, but each pulse provides a full spectrum. It turns out that the hardware needed to implement a pulsed ultrasonic spectrometer is actually less complex than in the case of the other two excitation methods described above, so this approach is very attractive from an economics perspective as well.

8.4.3   A Practical Ultrasonic Spectrometer

The author has built a series of practical ultrasonic spectrometers according to the design philosophy described above and demonstrated their use in a number of industrial processes; additional details concerning those applications are included in section 8.4.5. Figure 8.21 shows a picture of one such sensor, which was designed to measure PSD in a slurry process; the picture shows the flow cell assembly (called a *spool piece*) that mounts directly in the pipeline. The spool piece has the same inner diameter as the pipeline, and the transducers are mounted in the side of it in such a way that their ends

FIGURE 8.21 A full-bore ultrasonic sensor for slurry applications. The two transducers are mounted in the side of the spool piece, through which the calendar in the background can be seen. The power regulator, clock circuit, and pulser are mounted on the top plate; the preamplifier and a fiber optic link are located on the bottom plate.

are essentially flush with the inner wall. This "full bore" design imposes no restriction on the cross-sectional area of the pipe, so there is no additional pressure drop caused by the sensor. There are no moving parts in this design, and all exposed surfaces are made of either 316 stainless steel or polished silica.

The block diagram of this system is shown in Figure 8.22. The system clock is a simple oscillator (running at about 100 Hz) that simultaneously triggers the pulse generator and the high-speed analog-to-digital converter, which begins recording the received signal. The pulser generates a single, very short electrical pulse, causing the transducer to emit a broadband ultrasonic pulse. The bandwidth of the ultrasonic pulse itself is of course determined by the input power spectrum of the excitation pulse and the

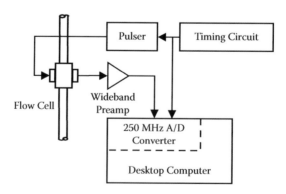

FIGURE 8.22 A block diagram of a generic pulsed system. From Scott, 1998; used with permission.

bandwidth of the transducer; other transducer effects are described in section 10.2.4. In the sensor shown in Figure 8.21, the excitation pulse width is approximately 10 ns and the ultrasonic pulse contains component frequencies ranging from about 10 MHz to over 30 MHz. The frequency range is picked to provide the most sensitivity to particle size in a particular process, and it is determined by the choice of transducer and the shape of the excitation pulse.

After the pulse propagates through the slurry, it is detected with the other transducer; the resulting electrical signal is amplified and finally digitized by a high-speed (in this case, 250 MHz) analog-to-digital converter. The digitized signal is the time domain representation of the ultrasonic pulse. Meanwhile, the wall of the flow cell reflects the ultrasonic pulse, thus forming an echo that reverberates through the cell for several transit times. These echoes limit the possible measurement repetition rate to less than 10 kHz.

The computer then selects the portion of the digitized signal containing the first received pulse in order to ignore the echoes; this process is called *gating*. The delay between the triggering of the excitation pulse and the arrival of the ultrasonic wave at the receiver, minus the delay that occurs within the transducers, is called the *time of flight* of the ultrasonic pulse. If the acoustic path length is known, then the group velocity (average sound speed of the pulse) can be measured. Since the sound speed is sensitive to temperature changes, the time of flight can be used to monitor the temperature within the slurry. The computer automatically adjusts the gating of the pulse according to changes in the time of flight so that the gate always contains the entire pulse.

A Fourier transform (implemented as the well-known *Fast Fourier Transform*) converts the gated time domain signal into the frequency domain. At this point the data is a discrete sampling of the amplitude and phase factor (represented as a complex number) of the frequency components comprising the received ultrasonic pulse. The transmission spectrum $T(f)$ is calculated by taking the common logarithm of the magnitude and multiplying it by 20; the data is therefore on a logarithmic scale which is equivalent to a decibel (dB) level, except that the reference is arbitrary.[8] (Computationally this treatment is equivalent to setting the reference value to 1, but so far we have not specified a scale factor for attenuation.) A typical spectrum obtained for the pulse excitation signal of Figure 8.20c is shown in Figure 8.23.

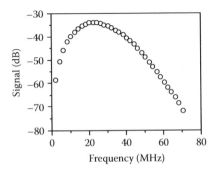

FIGURE 8.23 A digitized spectrum of the ultrasonic pulse shown in Figure 8.20c.

It should be noted that the spectrum in Figure 8.23 is not yet an attenuation spectrum, because it records the amount of sound that has propagated through the suspension or slurry, not the amount of sound that has been attenuated. Moreover, the spectrum of the digitized pulse depends not only on the frequency-dependent attenuation within the slurry, but also on the receiving transducer and the frequency response of the amplifier and digitizer. Taken together, these electrical components act as an amplifier with a frequency-dependent gain factor $G(f)$, which can be viewed as the intrinsic response of the system. Note that if the total system gain is less than 1, as it generally is, the signal level decreases. As suggested by Figure 8.24, to a first approximation the gain $G(f)$ is simply the product of the gain of each stage:

$$G(f)=G_1(f)\cdot G_2(f)\cdot G_3(f)\cdot G_4(f)\cdot G_5(f)\cdot G_6(f)\cdot G_7(f)\cdot G_8(f)\cdot G_9(f) \qquad (8.16)$$

Mechanical loading effects (which depend on the density of the slurry) can change the frequency response of the transducers. Also there is a small insertion loss at the transducer/slurry interface due to reflection of the ultrasound at the acoustic discontinuity. These factors can be influenced by the physical properties of the slurry, and the dashed

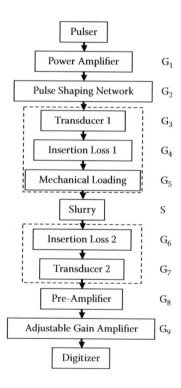

FIGURE 8.24 Factors contributing to the intrinsic response of a generic ultrasonic spectrometer; some may be omitted in a particular design.

lines around some components in Figure 8.24 indicate this dependence. The spectrum $T(f)$ measured by the system in Figure 8.23 is therefore given by

$$T(f) = 20\log_{10}[P(f) \cdot S(f) \cdot G(f)] \tag{8.17}$$

where $P(f)$ is the original pulse, $G(f)$ is given by equation 8.16, and $S(f)$ is the frequency-dependent attenuation due to the slurry. The intrinsic response of the sensor must therefore be considered in order to measure the attenuation of the slurry.

One method for addressing the intrinsic response issue is to measure a background spectrum $T_0(f)$ on the liquid phase of the slurry. In the case where $G(f)$ is approximately independent of the material in the flow cell, it follows that the difference between the background $T_0(f)$ and the sample $T(f)$ is the excess attenuation expressed in dB:

$$T_0(f) - T(f) = 20\log_{10}\left[\frac{P(f) \cdot S_0(f) \cdot G(f)}{P(f) \cdot S(f) \cdot G(f)}\right] = -20\log_{10}\left[\frac{S(f)}{S_0(f)}\right] \tag{8.18}$$

Since the attenuation coefficient $\alpha(f)$ has units of decibels per unit length, we divide this difference by the acoustic path length z (which is simply the spacing between the two transducer faces) to determine the attenuation spectrum:

$$\alpha(f) = \frac{T_0(f) - T(f)}{z} \tag{8.19}$$

The quantity $\alpha(f)$ given in equation 8.19 is the excess attenuation that is calculated by the ECAH theory. To the extent that the various gain factors in equation 8.16 do not change over time, the use of a background measurement is justified. However, if the transducer efficiency changes due to aging or a buildup of process material, this method will generate increasingly incorrect data. Therefore, the system should be checked periodically by remeasuring the background signal $T_0(f)$ or by measuring $\alpha(f)$ of a standard material for which the attenuation coefficient is known.

Another method for considering the intrinsic response of the ultrasonic spectrometer is to make measurements T_1 and T_2 at two separations z_1 and z_2 between the transducers. If the distances z_1 and z_2 are well inside the near field of the transducers, as is the case for relatively small transducer separation and high frequencies (above 30 MHz or so), then

$$\alpha(f) = \frac{T_1(f) - T_2(f)}{z_2 - z_1} \tag{8.20}$$

where it assumed that $z_1 < z_2$.[9] This method requires that one of the transducers be movable, and transducer alignment can be difficult to maintain. This added complication raises the cost and potentially decreases the reliability of the sensor, but the benefit is that a background measurement $T_0(f)$ is not needed. An example of a sensor that uses this approach is given in section 8.4.5.

Once the frequency dependent attenuation coefficient $\alpha(f)$ is measured, the appropriate ultrasonic theory is used to extract the PSD, as described in section 8.5. The inversion of ultrasonic data is essentially the same as that used in the case of static light scattering.

8.4.4   Validation of the Technique

Extensive experimental work has verified that the ultrasonic sensor described above can be used to determine PSD. Typical results are given in this section for the case of a dilute suspension of TiO_2 in water.

Validation of the ECAH Model

To illustrate the applicability of the ECAH model, ultrasonic attenuation measurements were made on dilute slurries made from commercial grades of TiO_2 powder. The samples were prepared in an aqueous solution of 0.1 wt% potassium pyrophosphate (which acts as a dispersant) and treated with high power ultrasound for two minutes.[10] Measured values of $\alpha(f)$ for two suspensions, each with a volume concentration of 0.5%, are shown in Figure 8.25. The expected attenuation was calculated from the ECAH model (which assumes particles of a single size) by integrating the attenuation over the known particle size distribution (see the inset in Figure 8.25) obtained with a Brookhaven X-ray disc centrifuge (XDC), which is briefly described in section 8.6 (Allen, 1992). The results of this "empirical calculation" are depicted as a line for each sample. The striking agreement

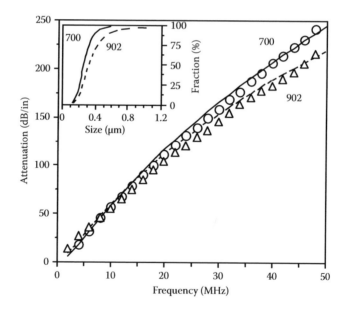

FIGURE 8.25 Attenuation versus frequency for two different size distributions of TiO_2. Points are measured spectra, and the lines depict forward calculations based on the A–H model. Inset: PSD obtained with a Brookhaven X-ray disc centrifuge. From Scott et al., 1995; used with permission.

between experiment and theory suggests that this model adequately predicts the outcome of the forward problem.

Two features of Figure 8.25 merit comment. First, the close agreement between experiment and theory suggests that this model adequately predicts the outcome of the forward problem for this material. The model slightly overestimates the observed excess attenuation by a few percent in the 20–30 MHz range, but the overall agreement is quite good considering that the model is an *a priori* calculation with no free parameters and 16 physical constants. Second, the PSD of the slurry clearly affects the ultrasonic attenuation, and the ECAH calculations correctly predict the effect. These observations suggest that the PSD can be measured on the basis of ultrasonic attenuation data.

Validation of the Inverse Algorithm

Based on the combination of a log-normal distribution with the ECAH model, the inversion algorithm outlined in section 8.5 was used to extract the PSD from the ultrasonic data obtained on a dilute suspension of DuPont Ti-Pure® R902 titanium dioxide. The result of the fit is shown as a line in Figure 8.26a, and the comparison between the

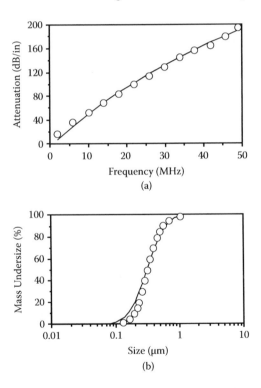

FIGURE 8.26 Ultrasonic data for R902 titanium dioxide: (a) results of the fitting procedure; the circles are ultrasonic data for the suspension, and the line is the best fit attenuation predicted by the ECAH model using the assumption of a log-normal size distribution; (b) particle size distribution; the line shows the PSD used in the calculation shown in (a), and the circles are the PSD as measured by the XDC. Adapted from Scott et al., 1995; used with permission.

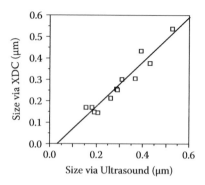

FIGURE 8.27 The particle size data measured with ultrasound shows very good agreement with XDC data.

corresponding log-normal distribution and the PSD measured by the XDC is given in Figure 8.26b. It should be noted that the mean size is accurately captured by the fit, but the reported width of the distribution is slightly broader than reported by the XDC. This result appears to be typical of the ultrasonic spectroscopy method.

The technique was also validated on four DuPont RPS Vantage® TiO$_2$ slurry products, which are highly concentrated (about 72 wt%) formulations for paper coating applications. Samples of several grades were diluted to 2 wt% and analyzed with the ultrasonic sensor and the XDC. Figure 8.27 compares the corresponding x_{16}, x_{50}, and x_{84} sizes measured by the two instruments. The linear regression, with an R^2 value of 0.928, indicates a very strong correlation between the two sets of measurements. The correspondence is not perfect; the line indicates a slight systematic difference of a few percent, which is due not only to errors in the physical parameters used in the ECAH model but also to the biases in the XDC and ultrasonic attenuation measurements. However, it is evident that ultrasonic spectroscopy gives reasonably accurate results for these submicron particles.

8.4.5   Applications

The author has applied the ultrasonic sensor described in the previous section to a variety of manufacturing processes, as summarized in Table 8.2. These processes are widespread throughout chemical industry, and most of them are not compatible with other online measurement techniques. A few of these applications are discussed below.

Monitoring and Control of Milling

Media mills are widely used throughout industry to disperse powders into a liquid and to reduce the particle size in slurries. A mill is essentially a grinding chamber where ceramic grinding beads (called grinding media) are agitated at high speeds. The slurry to be milled fills the space between the beads in the chamber, and the particles are broken when they are caught between the colliding beads. Some applications pump slurry through the mill only once, but often the material is recirculated through the mill until the desired particle size is reached.

TABLE 8.2 Examples of Various Applications of the In-process Ultrasonic Sensors Developed and Implemented by the Author

Application/Process	Segment	Goal
Grind-Point and Dispersion Monitor		
TiO2 high solids grinding	Coatings	Process development
CaCO3 grinding in DMAc	Materials	Process development
Organic pigments for inks	Printing	Product/process development
Production of dispersions for coatings	Coatings	Process monitor and control
Characterization of Droplet Size in Emulsions		
Production of Neoprene	Materials	Process development
High performance adhesive	Electronics	Product development
Characterization of Dispersion and Emulsion Stability		
Crop protection emulsions	Agriculture	Product development
Inks	Printing	Product development
Reaction Monitor		
Production of silver powder	Electronics	Process monitor and control

There are several economic benefits to online measurement of PSD in milling operations. If a material is milled too much, the slurry may become colloidally unstable, meaning that the particles start to agglomerate or flocculate; this degradation of the slurry requires corrective action to maintain product quality. Overmilling also wastes energy and causes unnecessary wear of the media and of the mill itself. Media mills represent a significant capital investment, so online measurements of PSD help to maximize asset productivity by indicating the completion of a mill run.

Previously, online measurements of PSD have relied on laser diffraction, which requires isokinetic sampling and dilution. An early application of the in-line ultrasonic instrument described above was to study the comminution (i.e., size reduction) of submicron particles of both organic and inorganic pigments. Figure 8.28 shows the decrease in median particle size during the milling of TiO_2 slurry in a semiworks milling loop (Scott et al., 1998). The concentration was rather high (75 wt%), and the acoustic path was so long (nearly 4 cm) that in order to make the measurement, transducers with a center frequency of only 0.5 MHz were used. The flow rate through the mill was fairly low, and the resulting residence time within the mill accounts for the delay between the start of milling and the point in time when the size started to decrease. The mill was turned off for about 10 minutes during the run, and the sensor indicated that the particle size was stable; when the mill was started up again, the recorded size started to drop. It is evident from the data that the milling limit was reached at 40 minutes.

Figure 8.29 shows the results of a similar test in which the material (a 10 wt% suspension of TiO_2) became overmilled. For this experiment, the sensor was installed in a

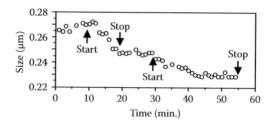

FIGURE 8.28 In-line real-time measurement of the median size of titanium dioxide at a concentration of 75 wt%. The mill was started and stopped at the times shown in the figure. Adapted from Scott et al., 1998; used with permission.

separate instrumentation loop that monitored the particle size of the suspension in the mill's reservoir, which was a small well-mixed tank. After an initial decrease, the apparent particle size (shown here as the mean size) started to increase even though the mill was running, and the increase continued for nearly an hour. The addition of a surfactant to the reservoir immediately caused the apparent mean size to decrease, but as the milling continued, the particle size grew again. Figure 8.29 is an example of the dynamic experimentation that would be impossible without an ultrasonic sensor.

A result from another comminution application is shown in Figure 8.30. In this case, the task was to grind calcium carbonate ($CaCO_3$) in dimethyl acetamide (DMAc) at about a 20 wt% concentration. The continuous attenuation at a single frequency (left ordinate scale) correlates very well with the median particle size (right ordinate scale) as measured offline via laser diffraction. Gaps in the ultrasonic data were caused by upsets in the mill when it was shut down in order to get a sample for the diffraction instrument. The direct correlation between particle size and attenuation evident in this figure suggests that in this case a simple calibration can be used to estimate particle size, thereby avoiding the expense of adapting the ECAH model to this material system. This approach must be used with great care, however, since the relation between size and attenuation is not monotonic over all particles sizes.

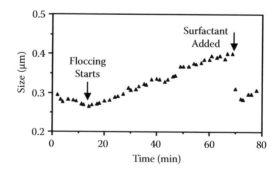

FIGURE 8.29 Data from a milling experiment in which the particle size starts to increase due to floccing; the flocs are broken up immediately upon the addition of surfactant. From Scott et al., 1998; used with permission.

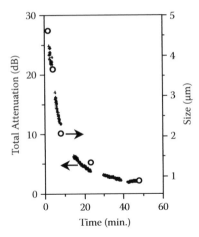

FIGURE 8.30 Data from a grinding study of CaCO₃ in DMAc suggests a direct correlation between the median size and the ultrasonic attenuation measured at a single frequency.

Figure 8.31 shows a picture of the sensor mounted in a milling loop at a paint plant, where it has been used to monitor the dispersion of an organic pigment in water, together with signal strength data from three separate production lots. The signal strength is the value of T (from equation 8.17) at 11 MHz; the scale has been offset slightly to set the noise floor at 0 dB. Two of the batches give nearly identical readings, while the third gives similar readings that are delayed in time. The signal strength is below the noise floor for the first three hours of milling, during which time the pigment is disagglomerated and reduced in size. Once the pigment is fully dispersed, the signal strength reaches a stable value. Note that the final reading is the same for all three batches. The rest of the

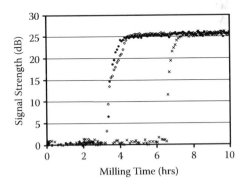

FIGURE 8.31 The sensor of Figure 8.21 is shown here in a milling loop at a pigment plant. The spool piece is mounted directly in a vertical pipe and is enclosed in a nitrogen-purged box (which is open in this photo). The chart shows the signal strength as a function of time for three production lots.

spectral data can be used to determine the absolute particle size, but for this application it is sufficient to know when the pigment is fully dispersed.

Droplet Size Measurements in Emulsions

Neoprene is an elastomer (synthetic rubber) used to manufacture a wide range of common items including mechanical belts, inner tubes, gloves, gaskets, garden hose, and plumbing fixtures. It is manufactured using an emulsion polymerization process in which an initiator is added to an emulsion of chloroprene in water. Variability in the emulsion affects the mechanical properties of the product, so control of the chloroprene droplet size distribution (DSD) is important. Measurement of the DSD is difficult because the emulsion is milky in appearance; moreover, chloroprene is a flammable solvent, so the instrument must be intrinsically safe.

The ultrasonic probe shown in Figure 8.32 was designed to operate in pulse echo mode so that measurements could be made from the top of a large process tank. It was assumed that a thin layer of material would build up on the transducer face and thereby change $T_0(f)$, so the probe was designed with a variable acoustic path length. The reflector in the probe can be moved by two miniature pneumatic cylinders to one of two positions, either 4 or 5 cm from the end of the transducer. The acoustic path for the reflected pulse is therefore either 8 or 10 cm, and equation 8.20 can be used to determine $\alpha(f)$. The transducer in this probe covers the 30–70 MHz frequency range, and during testing it was found that the ECAH model worked quite well for 50 volume percent emulsions of chloroprene.

In tests at the manufacturing plant, emulsion samples were prepared under a variety of operating conditions. An experiment was arranged so that droplet size measurements could be made simultaneously with two instruments: the ultrasonic sensor and the FBRM optical probe described in section 8.2.4. The results are compared in Figure 8.33, and the agreement is excellent ($R^2 = 0.991$) for droplets with diameters in the range of 10–13 μm. It was observed that as the droplet size decreased the difference between the two results widened due to systematic error, but the overall correlation remained quite strong ($R^2 > 0.99$). These results demonstrate that ultrasound can be used to measure DSD even under adverse conditions.

Precipitation

A unique application of ultrasonic spectroscopy is in the precipitation of silver particles used in the production of electronic pastes. In the manufacturing process, a large (2 kL) batch of

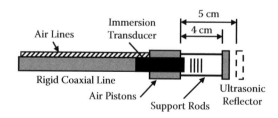

FIGURE 8.32 The ultrasonic probe used to measure the droplet size distribution of a chloroprene emulsion.

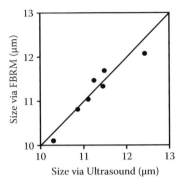

FIGURE 8.33 A comparison between the droplet size in an emulsion of chloroprene and water, as estimated by simultaneous ultrasonic and laser back-scatter measurements. The line shows the linear regression.

silver solution is prepared from silver nitrate and monoethanolamine (MEA) in a header tank. Silver clusters (20–50 nm in diameter) form in this reaction and grow slowly over time. After an aging period of many hours, the solution is reacted in a series of small batches with a reducing agent to precipitate silver particles, which are then filtered, washed, and dried for later use. Batches of silver are made as needed; sometimes the header tank is emptied within days, and sometimes it lasts for two weeks. The size of the precipitated particles is observed to depend upon the amount of aging, suggesting a dependence on the size of the silver complexes in solution. The silver solution itself is highly reactive, and in order to control the process an *in situ* measurement was needed.

Since the header tank already had a flange mounted near the bottom, it was decided to construct a probe that could be installed directly into the tank. This probe, depicted in Figure 8.34, features a pair of transducers mounted on stainless steel extension tubes that attach to the mounting flange. The transducers face each other, and the gap between them is fixed at 4 cm; the frequency range is 20–70 MHz. The electronics package is mounted on the flange and communicates with the host computer over a fiber optic link.

FIGURE 8.34 A photo of the ultrasonic spectroscopy probe used in silver powder production.

The transducers had to be modified to survive contact with the silver solution; it was found that the base of the transducer was nickel-plated brass, most of which dissolved after a few weeks of exposure (fortunately the transducers did not stop working). This problem was solved by coating the base of the transducers with plastic; the system subsequently survived a full year in the plant process without failure, and in fact it was still working well when the project ended.

Plant tests were conducted to study changes in the ultrasonic attenuation spectrum due to aging of the silver solution and to relate them to the silver particle size obtained after precipitation. The system was calibrated before and after the experiments using a water path measurement, and no drift was observed. (Even after a 12-month exposure to silver solution in the process, the water transmission measurement shifted by only a few decibels.) To demonstrate the concept, during the aging cycle samples of silver solution were periodically combined with reducing agent to precipitate silver particles. These particles were later imaged via scanning electron microscopy (SEM) and the resulting images were analyzed to determine the particle size; the mean size was determined from over 200 particles in each sample. After several hours, extra MEA was added to the silver solution in order to cause a step change in the cluster size and ultimately in the precipitated particle size.

The results are shown in Figure 8.35, where the attenuation at 50 MHz of the silver solution is shown as a line (scale on the left) and the size of the corresponding precipitated silver particles is indicated by filled circles (scale on the right). Typical SEM images of the precipitated silver are shown as insets; the scale bar is 1 μm in length. The aging time is shown on the horizontal axis. It is clear that when the silver complexes in the solution were altered by the addition of extra MEA, the size of the precipitated silver

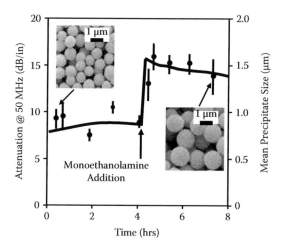

FIGURE 8.35 Correlation between the attenuation of ultrasound in the silver solution and the size of particles precipitated from that solution. The inset photos are scanning electron microscope (SEM) pictures of the precipitated particles after they have been washed and dried. The nanoclusters in the silver solution and the precipitated particles both increase in size after the addition of monethanolamine.

particles was also affected. A gradual decrease of attenuation in the silver solution and of precipitate particle size was also observed following MEA addition, suggesting that the chemical reactions in the tank had not yet reached an equilibrium. In effect, it is possible to predict the size of the precipitates based on the size of the nanoparticles in the silver solution, and this information can be used to take corrective action before unacceptable powder is produced (Scott, 2006). This ability to monitor the solution aging process enables feed-forward control of final particle size.

8.5 Appendix on Recovery of the PSD

This appendix presents a generic algorithm for extracting the PSD from measurements of optical or ultrasonic scattering. The input data may be a diffraction pattern, ultrasonic attenuation spectrum, phase velocity spectrum, or virtually an other physically observable vector quantity that contains information about particle size. This operation is a typical example of an inverse problem.

Given a mathematical model of a physical measurement system, inverse problems estimate the input parameters of the model that would be required to produce the observed output data. Inverse problems are naturally the opposite of the forward problem, where the model input parameters are already known and the outcome is predicted for comparison with observation. The observed data must reflect specific properties (in this case, the PSD) of the system studied. Obviously, the physical model used to describe the measurement process must an accurate representation of that process.

For simplicity we will assume that the suspension is dilute, so at least locally the ultrasound interacts with only a single particle; in this case the observed signal (represented by the vector **s**) is formed by the superposition of individual, uncorrelated events, and the problem is linear in nature:

$$\mathbf{s} = \mathbf{Mq} \qquad (8.21)$$

where M is a linear operator that represents the model, and q is a vector containing the differential PSD, $\{q_i\}$. Notice that all of the physics is contained in the matrix $\mathbf{M}$. Given the measurement vector **s**, one might be tempted to use an inverse matrix $\mathbf{M}^{-1}$ to determine the particle size **q** directly, as in

$$\mathbf{q} = \mathbf{M}^{-1}\mathbf{s}. \qquad (8.22)$$

However, this approach often yields results that physically unacceptable, for example negative size fractions. Direct inversion is particularly risky when the determinant of M is close to zero. Furthermore, the information contained in **s** can easily be distorted by measurement errors, leading to highly oscillatory solutions that are physically irrelevant.

Generally, inverse problems are said to be ill-posed because they violate one or more of Hadamard's (1923) three postulates of "well-posedness":

1. For all possible measurements *s*, a solution *q* exists
2. For a given *s*, this solution is unique
3. The solution depends continuously on the data

The second condition usually presents the biggest challenge, since a whole family of solutions is often possible. It should also be noted that even if *s* were measured with complete accuracy, it may not convey adequate information about the PSD. In particular, premature truncation of the attenuation spectrum or diffraction pattern dramatically reduces the information content.

A simple but robust approach to inversion is to use a least-squares fitting algorithm that minimizes the distance Δ, where

$$\Delta = \| \mathbf{Mq} - \mathbf{s} \|. \tag{8.23}$$

Using a method called *numerical quadrature*, the particles in the suspension are divided into N discrete size classes and the observed data is written as a set of linear equations (Twomey, 1963):

$$\mathbf{s} = \mathbf{Mq} + \mathbf{w} \tag{8.24}$$

where *s* is the vector containing the observed data (optical or ultrasonic), $\mathbf{M}$ is the matrix form of the discretized model, $\mathbf{q}$ is the solution vector representing the unknown size distribution, and $\mathbf{w}$ represents the vector of experimental measurement errors in *s*. Possible sources of measurement error include:

- temperature dependence of the sensor
- misalignment of the detectors
- incorrect measurement of the intrinsic response or signal baseline

Systematic errors may also originate from use of an incorrect physical model or of inaccurate physical constants within that model.

Regularization, an operation that builds Hadamard's third condition into the problem, can be provided by forcing a connection between individual size fractions through the assumption of a specific PSD function, $p(x)$. This operation may be essential when inverting measurement data that is relatively smooth. For example, the predicted ultrasonic attenuation spectra of Figure 8.18 and the observed spectra of Figure 8.25 and Figure 8.26 could all be described (within experimental error) by a quadratic function, which has only three parameters. Thus, it would be difficult (and perhaps foolish) to extract more than three independent parameters from such data. In many industrial applications, a suitable description of the PSD is the log-normal distribution:

$$p(x) = \frac{1}{\sqrt{2\pi}(\ln\sigma)x} \exp\left[-\frac{(\ln x - \ln x_{50})^2}{2(\ln\sigma)^2} \right] \tag{8.25}$$

where σ is the geometric standard deviation of the distribution and x_{50} is the median size as defined earlier. If we denote c as the total volume fraction of solids in the suspension, then it follows that the differential size distribution $q(x)$ can be written as

$$q(x) = c \cdot p(x) \tag{8.26}$$

This choice of PSD greatly reduces the number of parameters to be estimated to only three: concentration c, median size x_{50}, and geometric standard deviation σ. Of course, the assumption of a particular shape of $p(x)$ can lead to erroneous results if the true PSD does not conform to it.

A standard fitting method, depicted in Figure 8.36, is used to estimate the PSD. This approach avoids the mathematical problems associated with direct inversion of near-singular matrices:

1. Measure the scattering data s.
2. Make an initial guess of the PSD, $\hat{\mathbf{q}}$. If a particular shape is assumed for the PSD, this step is equivalent to making an initial guess for the PSD parameters and calculating the PSD. In the case of a log-normal distribution, the parameters are c, x_{50}, and σ, and equations 8.25 and 8.26 are used to calculate $\hat{\mathbf{q}}$.
3. The expected data $\hat{s}$ is predicted by equation 8.21.
4. Calculate the distance Δ between $\hat{s}$ and s by using equation 8.23.
5. Update the estimate of $\hat{\mathbf{q}}$ (or the parameters used to calculate it) and iterate until Δ becomes sufficiently small.
6. The final $\hat{\mathbf{q}}$ is the estimated PSD.

The validity of the PSD produced by this algorithm can be assessed by examining the fit of the model to the observed data. The residual vector r, whose components represent

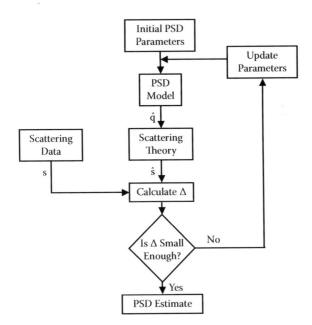

FIGURE 8.36 An algorithm for extracting PSD from measurements of optical or ultrasonic scattering.

the difference between the predicted and observed data at each angle or frequency, is weighted with the experimental error σ_e:

$$\mathbf{r} = \frac{\mathbf{s} - \hat{\mathbf{s}}}{\sigma_e} \tag{8.27}$$

The reduced chi-square value is calculated from the sum of the squares of the residuals:

$$\chi^2_{red} = \frac{\mathbf{r} \cdot \mathbf{r}}{(m - N)} \tag{8.28}$$

where m is the size of the measurement vector (i.e., the number of angles or frequencies comprising *s*) and N is the number of size classes or parameters to be determined. The value of the reduced chi-square (χ^2_{red}) is a measure of the goodness of fit and thus indicates whether the inversion has been successful. When $(m-N)$ is large enough, χ^2_{red} is normally distributed with mean $(m-N)$ and variance $2(m-N)$. If χ^2_{red} is unsatisfactorily large, then the cause for the misfit should be analyzed. It is generally held that a model is acceptable if χ^2_{red} approaches unity.

Assuming the model errors are normally distributed, an estimate for the asymptotic standard deviation (or standard error, SE) follows from

$$SE = \sqrt{\chi^2_{red}(\mathbf{M}^T\mathbf{M})^{-1}} \tag{8.29}$$

where $(\mathbf{M}^T\mathbf{M})^{-1}$ is the variance-covariance matrix with dimensions $(N \times N)$.

Next we can compute the 95% confidence intervals for the model parameters as $\pm SE \cdot t_{\alpha/2}(m\text{-}N)$, where $t_{\alpha/2}$ represents the student *t*-distribution and α is 0.05. For a complete assessment of the susceptibility of the model parameters to perturbations, the best estimates of the parameters x_{50}, σ, and c are individually varied, consequently raising the value of χ^2_{red} (Figure 8.37). It can be proven that the χ^2_{red} follows a parabola, and some fitting routines make use of this fact. Pinpointing the confidence boundaries for the parameters involves varying one parameter at a time until the limits are reached. The values for these limits follow directly from the *F*-distribution. The shape of the parabola varies for

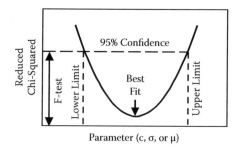

FIGURE 8.37 Confidence intervals for estimated parameters produced by a fit to the observed data.

all parameters, directly indicating the degree to which one parameter affects the fit. For example, a steep parabola for a particular parameter indicates a strong dependency, suggesting that the estimated value is well-defined. The results of this analysis are confidence intervals for x_{50} and σ, which allow discrimination of significant changes in the PSD.

8.6 Appendix on the X-ray Disc Centrifuge

The XDC was used to generate the comparison PSD data shown in Figure 8.26 and Figure 8.27. The technique is introduced briefly here for completeness, but the interested reader should consult Allen (1997, chapters 6–8) for additional details.

The XDC is based on the sedimentation technique, which measures the settling speed of particles as they fall through a liquid. Particles fall at a terminal velocity, where the gravitational force on the particles is balanced by the viscous drag force exerted by the liquid. For spherical particles with a very low Reynolds number, the relationship between the settling velocity v and the particle diameter x is derived from Stokes's Law:

$$x = \sqrt{\frac{18\eta v}{(\rho - \rho_0)g}} \tag{8.30}$$

where η is the viscosity of the fluid, g is the acceleration due to gravity, ρ is the density of the particle, and ρ_0 is the density of the fluid.

Fine particles tend to settle slowly, and equation 8.30 breaks down for very small particles where Brownian diffusion becomes significant. In such cases, diffusion can be overcome by placing the suspension in a centrifuge that rotates at a high speed. The centrifugal acceleration effectively increases the "gravitational" factor and causes the particles to settle relatively quickly. In the XDC, the suspension is placed in a hollow disc; as the disc spins, the solids move toward the outer rim. The motion of the solids is monitored with a collimated X-ray beam that passes through the disc and the suspension; the attenuation of the beam increases in proportion to the concentration of particles. Thus, the fractional increase in X-ray attenuation as a function of time can be converted to a fractional increase of mass as a function of velocity. To a first approximation the velocity can be cast in terms of Stokes's diameter via

$$x = \sqrt{\frac{18\eta v}{(\rho - \rho_0)\omega^2 R}} \tag{8.31}$$

where R is the radial distance from the axis of rotation to the X-ray beam and ω is the angular velocity in radians per second. The effect due to gravity is generally quite small compared to the centrifugal force.

Suggested Reading

Allen, T. (1997). *Particle Size Measurement,* Fifth edition, *Vol. 1.* London: Chapman & Hall.

Bohren, C.F. and Huffman, D.R. (1983). *Absorption and Scattering of Light by Small Particles.* New York: Wiley.

Dukhin, A.S. and Goetz, P.J. (2002). *Ultrasound for Characterizing Colloids*. Amsterdam: Elsevier.

Ishimaru, A. (1997). *Wave Propagation and Scattering in Random Media*. New York: IEEE Press.

Kerker, M. (1969). *The Scattering of Light and Other Electromagnetic Radiation*. New York: Academic Press.

Povey, M.J.W. (1997). *Ultrasonic Techniques for Fluids Characterization*. San Diego, CA: Academic Press.

9

Process Imaging

9.1 Introduction

Measurement and control technology can simplify the operation of process equipment, improve product quality and asset productivity, and minimize waste by increasing first-pass yield. Thus, there are real economic incentives for improving the control of these processes. In the past, the feedback used for industrial control systems was based on scalar quantities such as temperature and pressure, which were measured at single points in the process. Due to the increasingly stringent demands on the quality of products produced by increasingly complex systems, more sophisticated measurements are needed to control manufacturing processes. Two-dimensional (2D) and sometimes three-dimensional (3D) information is needed to optimize processes such as casting of polymer film and mixing chemical components. Process imaging techniques provide this higher dimensionality.

The term *process imaging* refers to the combination of image-based sensors, data processing concepts, and display methods used to produce information about the internal state of industrial processes. This data is used for research and development of new processes and products, process monitoring, and process control. One can characterize the physicochemical state of an industrial process based on spatio-temporal patterns in planar (2D) or volume (3D) images. These images can be obtained directly with a camera, or indirectly via tomographic reconstruction, as explained in section 9.3. A systematic review of this fascinating and relatively new field is provided by Scott and McCann (2005).

Direct process-imaging techniques record images of material inside process vessels with a video camera, laser scanner, or other suitable imaging system. The images were traditionally recorded on photographic film, but charge-coupled device (CCD) video cameras and other electronic sensors have supplanted the use of film. Sometimes the scenes are invisible to the human eye, as for instance in infrared or X-ray imaging applications; in those cases, specialized equipment is necessary to capture the image. Both infrared and X-ray radiation can pass through many process materials and reveal something about the interior structure or density variations. X-ray imaging in particular is used to see inside materials and processes. When visible light is used to illuminate opaque process material, one sees only the exterior, yet even visual appearance may be sufficient to convey useful information about the state of the process.

Tomographic imaging is indirect in the sense that the image must be mathematically reconstructed from a set of information recorded by an array of sensors situated around the boundary of the measurement subject. Raw tomographic sensor data is not an image in the usual sense, but the reconstructed image is a cross-sectional 2D or 3D view of the interior of the subject. Tomography has become familiar through applications in the field of diagnostic medical imaging of computed tomography or nuclear magnetic resonance imaging scanners. The application of this technology to industrial processes is called *process tomography* (see Scott & McCann, 2005; Scott & Williams, 1995; Williams & Beck, 1995). These techniques provide unique information about the internal state of the industrial process.

Regardless of how it is produced, a process image is only a means to an end: the information in the image must be extracted, interpreted, and put to use. These operations are based on computational techniques developed in the field of digital image processing (see Castleman, 1996; Rosenfeld & Kak, 1982). Digital images are simply large arrays of numbers that represent picture elements (pixels) of various levels of gray or color, so image processing is perfectly well suited for computer implementation. Image enhancement techniques are sometimes used to improve clarity or to accentuate certain features within the image. The features of an image are generally regions of pixels that are darker or brighter than surrounding pixels and that are spatially correlated with each other. After the features have been recognized, their position, size, shape, color, or texture is analyzed and recorded. These features contain the desired information about the industrial process, so the interpretation of the image is based on an analysis of the extracted features. This analysis is based on a mathematical model (usually a simplified one) of the sensor. Finally, the interpretation of the image is provided to the system operator or process controller for further action.

The foregoing explanation has been necessarily vague, because the details involved in process imaging are rather application-specific. The case studies presented in this chapter illustrate the application of imaging technology in three industrial unit operations, and they should clarify the concepts involved. Several other process imaging applications related to polymers and composite materials are discussed in chapter 11.

9.2   Direct Imaging

With the arrival of inexpensive electronic cameras and video recorders, the application of direct imaging technology has become both widespread and well known. The basic sensor used in direct imaging is the CCD camera chip described in section 5.2. Sometimes the light level is very low or the radiation is invisible, so an intensifier (like that described in section 6.4.2) is used.

Access and lighting are two critical aspects involved in direct imaging of industrial processes. Optical access to the relevant part of the process must be available in order to record a visible image. Fiber optic bundles that maintain the spatial relationship among individual pixels are widely available and are quite useful for viewing the interior of a process vessel or pipe; an example application of these imaging bundles is described in section 11.2. Visible lighting can be provided by a variety of sources, including light-emitting diodes, incandescent or fluorescent lamps, lasers, or strobe lights (flash lamps). Infrared lighting can be provided by heat lamps, resistive heaters, or (more likely) by the process itself. Fiber optic cables can be used to carry visible or infrared light from the source to the area of interest inside the process. X-rays for imaging applications are generated by a high-voltage electron tube as explained in section 6.3.2. The examples described below and in chapter 11 demonstrate how some of these components are integrated to provide process images.

9.3 Tomographic Imaging

Tomography reconstructs cross-sectional images that show the distribution or homogeneity of material in a chemical process. To examine how it works, we start by considering the transmission of parallel X-rays through a pipe (Figure 9.1) which is half-filled with a homogeneous process material that attenuates the radiation. A real X-ray source emits divergent rays, but the parallel ray geometry shown in Figure 9.1 illustrates the basic concept. If a detector array is placed on the other side of the pipe, it measures the intensity of the transmitted beam along each ray from source to detector. The intensity $I(x'; \varphi)$ is a function of the position x' across the detector and the orientation angle φ (which is treated as a parameter). If this angle changes, then the intensity pattern also changes, as shown in Figure 9.1. In the case of parallel X-rays, the intensity is

$$I(x'; \varphi) = I_0 \exp[-\int \mu(x, y) dy] \qquad (9.1)$$

where I_0 is the intensity of the incident radiation and $\mu(x, y)$ is the position-dependent linear attenuation coefficient of the material. Here the primed coordinates x' and

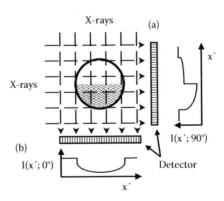

FIGURE 9.1 Observed intensity profiles of parallel X-rays after they pass through a half-filled pipe: (a) intensity of horizontal rays, $I(x'; 90°)$; (b) intensity of vertical rays, $I(x'; 0°)$.

y' are rotated with respect to the body-fixed coordinates x and y via an orthogonal transformation:

$$\begin{bmatrix} x' \\ y' \end{bmatrix} = \begin{bmatrix} \cos\varphi & -\sin\varphi \\ \sin\varphi & \cos\varphi \end{bmatrix} \cdot \begin{bmatrix} x \\ y \end{bmatrix} \tag{9.2}$$

The orientation angle φ is therefore the angle between the x-axis and the x'-axis. Note that the integration in equation 9.1 reduces the dimensionality of the information.

In X-ray tomography, the intensity data described by equation 9.1 are linearized by taking the natural logarithm with respect to the incident intensity; this data is called the *projection*, $P(x'; \varphi)$:

$$P(x';\varphi) = -\ln\left[\frac{I(x';\varphi)}{I_0} \right] = \int \mu(x,y)dy \tag{9.3}$$

The term *projection* is used to remind us that a lump of dense material sitting between the source and the detector at a location (x_0, y_0) casts a shadow on the detector at position x'_0, which is determined by equation 9.2. This position x'_0 changes with φ, and by noting the phase and amplitude of the function $x'_0(\varphi)$, we can locate the original position (x_0, y_0). It should be noted that the projections generated by a real X-ray source differ from equation 9.3 due to the divergence of the rays.

A complete set of projections can be obtained by measuring $P(x'; \varphi)$ for all values of φ. This set of projections constitutes the two-dimensional Radon transform, R. The foundation of tomographic reconstruction was laid in 1917 by Radon, who proved that, given the transformation R, an inverse transform exists such that

$$f = R^{-1}[R[f]] \tag{9.4}$$

for any arbitrary function f. His proof demonstrates that an N-dimensional subject can be reconstructed from an infinite number of $(N-1)$-dimensional projections. In tomography, the function f is the cross-sectional image $f(x, y)$ through an object or the volume image $f(x, y, z)$ of the interior of an object. Note that in any physical implementation only a finite number of projections can be measured, so the reconstruction is expected to contain a certain amount of error (Herman, 1980).

In Figure 9.2, the pipe has been filled with a mixture of two component materials. X-ray projections, calculated from the measured intensity data using equation 9.3, are shown for the extreme cases of (a) complete segregation and (b) complete mixing. The vertical projections $P(x', 0)$ for the two cases are nearly identical, but the difference between the horizontal projections $P(x', 90°)$ enables us to distinguish between a pipe in which the components are completely mixed, and one in which the components are completely segregated. In general, many projections are needed to determine the shape of objects that might be in the pipe, such as lumps of undispersed solids. The task of tomographic reconstruction is to determine the material distribution that would yield the observed projections, and the output is typically shown as a cross-sectional image. Tomographic reconstruction is an ill-posed problem, so it faces the same mathematical

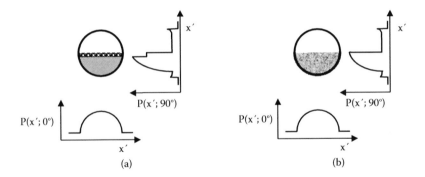

FIGURE 9.2 Horizontal and vertical X-ray projections of a pipe containing two components: (a) projections observed when the components are segregated; (b) projections observed when the two components are well mixed.

difficulties that were discussed in chapter 8 in conjunction with the determination of particle size distribution from ultrasonic spectra.

One way to reconstruct an image of the original object is to *back project* the value of $P(x'; \varphi)$ at each projected point x' in the measurement along the direction of y' (at right angles to the sensor) and to sum all the back projections together over all projection angles:

$$f_{FB}(x,y) = \int_0^\pi P(x\cos\varphi + y\sin\varphi, \varphi)d\varphi \qquad (9.5)$$

This technique of linear back projection is illustrated in Figure 9.3. Each one of the projections of a point object (Figure 9.3a) is a delta function (Figure 9.3b), which has a value of zero everywhere except a single point. The back projection of a delta function describes a line, and the integration of these lines over many angles is represented in Figure 9.3c. The final reconstructed image of the point object of Figure 9.3a (computed according to equation 9.5) is shown in Figure 9.3d, which is the point spread function for this type of reconstruction. Note that a different reconstruction formula must be used for divergent rays (Feldkamp et al., 1984).

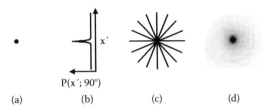

FIGURE 9.3 An example of linear back projection: (a) a point object has a projection (b) that is a delta function; the back projection which is a line; (c) the back projections from several orientation angles is integrated to form (d) the reconstructed image of the original point. Adapted from Hoyle et al., (2005).

It is evident that images tend to be blurred in this method of reconstruction. Due to the linearity of the imaging process, the reconstructed image of a more complicated object will simply be the superposition of a large number of blurred points. Although the point-spread function for back projected images is known to have a relatively long-range $1/r$ dependence (where r is the distance from the point), a convolution (filtering operation) performed on the projections before reconstruction improves the image clarity (Barrett & Swindell, 1977; Rosenfeld & Kak, 1982). This approach is the basis of the filtered back projection method, which is widely used.

An alternative reconstruction technique is to use an iterative algorithm, just as in the case of particle size measurement. Iterative techniques solve the inversion problem by comparing the projections of an assumed solution to the projections actually measured, adjusting the assumed solution for a better fit, and repeating the process until the solution converges. A finite element method is often employed for the forward calculations in such algorithms.

Tomographic techniques are not limited to the use of X-rays; in fact, much of the research effort in the area of process tomography has focused on measurements of electrical characteristics, such as capacitance and impedance. Electrical tomography techniques are interesting because they are fast and relatively inexpensive. *Electrical capacitance tomography* (ECT) is used to measure the capacitance between pairs of electrodes mounted on the perimeter of the pipe or vessel (Figure 9.4a). The capacitance between two electrodes is proportional to the average dielectric constant of the material between them, so capacitance values can be used as projection data. By measuring the capacitance between every electrode pair and back projecting the measurements along the field lines shown in Figure 9.4a, one can reconstruct a map of the dielectric constant within the process. Since the dielectric constant (i.e., relative permittivity) is a material

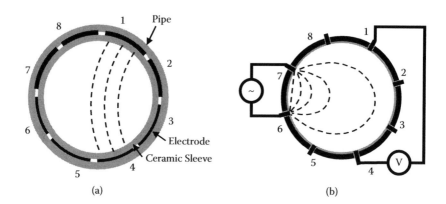

(a)   (b)

FIGURE 9.4 Electrical tomography in pipes: (a) ECT in an eight-electrode system, depicting idealized electric field lines between electrodes 1 and 4; the capacitance is measured sequentially between every pair of electrodes (1–2, 2–3, 1–3, etc.) and this data is subsequently back projected or otherwise processed to form a cross-sectional image in the plane of the electrodes; (b) ERT in an eight-electrode system, showing the injection of current between electrodes 6 and 7 and the measurement of voltage between electrodes 1 and 4.

property, this image provides information about solids distribution, mixing, or even the stage of chemical reactions taking place within the process.

Tomographic images of conductivity in electrically conductive materials are generated from resistance or impedance data by using *electrical resistance tomography* (ERT) or *electrical impedance tomography*. As shown in Figure 9.4b, these methods often use a four-wire measurement in which alternating current is injected between two electrodes while the resulting voltage drops between the other pairs of electrodes are measured. This approach reduces the undesirable influence of electrode resistance in cases where significant current must be injected to produce an adequate signal. Cross-sectional images are produced by back projection or finite element calculation. These images provide information about mixing kinetics and the dissolution of solids in a process.

For certain applications, tomography can also be accomplished with optical measurements. Many chemical compounds absorb ultraviolet or visible radiation and subsequently fluoresce (emit photons) at specific wavelengths, due to electronic transitions between quantized energy states. The spectrum of the emitted light is therefore characteristic of the molecular species present. The spatial distribution of selected chemical compounds can be determined by illuminating the process stream with beams of collimated light and measuring the emission intensity at specific wavelengths along those beams. An optoelectronic tomography system based on this concept has generated images that show the distribution of hydrocarbons in an experimental internal combustion engine (Sick & McCann, 2005).

9.4   Case Study: Crystallization

The morphology of product crystals is an important characteristic that can seriously affect the operation of centrifuges and dryers, not to mention handling characteristics such as flowability. Crystal morphology is influenced by a number of factors including vessel design, solids concentration, and temperature gradients. The traditional method of measuring particle morphology is to grab samples from the process and to inspect them under a microscope. However, in many cases the conventional method of removing crystals from the process does not preserve the crystal morphology. Removing crystals from the mother liquor causes additional growth and leaves fine particles stuck to larger particles. These effects drastically change the particle surface, making a representative sample difficult to obtain.

The problem of sampling can be solved with an in-line camera probe that can image and characterize the product crystals in the slurry exiting a full-scale industrial crystallizer (Scott et al., 1998). Many suitable industrial cameras are commercially available now, but a decade or two ago it was necessary to build one's own system, and even today certain applications still require a customized design.[1] A simple camera built for installation in an industrial crystallizer is shown in Figure 9.5. This camera is based on an industrial borescope (Schott model 10RS455D) used for inspecting the interiors of closed vessels. The camera probe is a 2.5-cm diameter stainless steel tube that protrudes into the process stream through a ball valve. A sapphire window at the end of the probe provides optical access to particulate material inside the process. Flashes from a

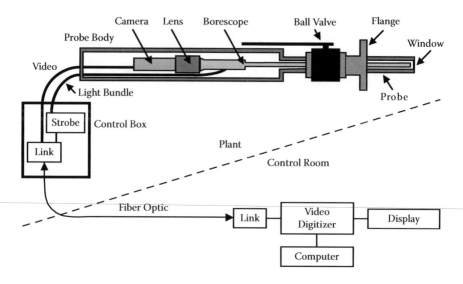

FIGURE 9.5 A camera probe for monitoring industrial crystallizers.

modified strobe light (EG&G Electro-Optics model MVS-2601) that is mounted in a nearby control box "freeze" the motion of the moving particles. Thus, particles that are flowing at velocities up to a few meters per second can be viewed without significant blurring of the image. Light is carried from the strobe to the probe window via a fiber optics bundle, and the image is relayed to a CCD camera by the borescope optics. The video signal is transmitted to a frame digitizer via a fiber optic link (Fiber Options models 215D and 110V), so the computer and operator interface can be mounted in any convenient location, such as the plant's control room.

This camera probe has been used at production sites to monitor the crystal morphology as part of a process-optimization program. The instrument was designed to be compatible with standard 1-inch (2.54 cm) ball valves that were already installed at various locations in the crystallization plants, so the camera can easily be moved from one location to another. Figure 9.6 shows an early in-line image taken during start-up of a crystallizer. The crystal edges have been rounded by attrition occurring within the crystallizer. Other crystal features that have been observed with this camera probe include crystal twinning, inclusions, and relative transparency (an indication of crystal purity). These observations are useful in diagnosing and correcting problems in the operation of the crystallizers.

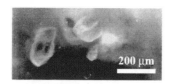

FIGURE 9.6 In-line image of crystals produced in a plant during start-up. From Scott et al., 1998; used with permission.

9.5   Case Study: Granulation

The purpose of a granulation process is to transform powder (i.e., fine particles) or mixtures of different powders into larger particles that have a desired size distribution and bulk density. The optimal size of the granules is on the order of 1–2 mm. Compared to the powders from which they are made, granules are much more convenient to transport, store, measure, and apply; therefore many intermediate chemicals and consumer products are granulated. Typical formation processes are extrusion from a paste made from the powder, and pan or fluidized bed granulation, in which a binder solution is sprayed onto the powder while it is agitated (Iveson et al., 2001). The size and shape of the resulting granules vary within a batch of product, so these characteristics are described by distributions rather than single numbers.

Uniformity in the product is highly desirable because large granules tend to rise to the top during material handling and transportation, and this segregation introduces variability in dissolution rate and other undesirable side effects in the product's end use. To obtain product uniformity, it is necessary to control feed rates, pan rotation speed, temperature, and other variables in the granulation process so that the size and shape distributions remain approximately constant. The control scheme must include an automatic measurement system capable of providing size and shape distributions every few minutes.

Extruded granules are difficult to measure with conventional instruments, such as those based on laser diffraction, due to their cylindrical shape and relatively large size (several millimeters in length). The most effective solution to this measurement problem is to extract the necessary information from images of the product material; an instrument based on this concept was developed at DuPont in the mid-1990s to control a granulation process (Scott et al., 2001). This instrument, which is described below, acquires images of the granules as they slide down an inclined plane and determines the length and width of each one (Sunshine et al., 2005). Granules can be measured at a rate of about 1000 per minute, so a complete size distribution with good counting statistics can be determined in a few minutes. Since the actual production rate is much higher than the measurement rate, this instrument is used to monitor a side-stream of product. In recent years several vendors have independently introduced similar commercial instruments.[2]

In this instrument (Figure 9.7) the granules are transported by a vibratory feeder from a small hopper to an inclined plane, down which they roll or slide under the influence of gravity. The feed rate is kept low so that only two dozen granules per second are tipped onto the slide, which is a smooth black surface inclined about 45 degrees from horizontal. The granules tend to spread apart from one another as they descend, which greatly simplifies the task of extracting information about their size and shape from the images. Some products generate a little bit of dust as they are conveyed; this dust tends to accumulate on the slide and interfere with the images, so the instrument periodically cleans the slide by releasing a blast of compressed air across it. A bright incandescent lamp illuminates the slide continuously, and a video camera records images of the granules. The electronic shutter speed of the camera must be fast enough (on the order of 100 μs) to avoid image blurring due to the motion of the granules. A computer analyzes

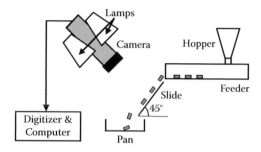

FIGURE 9.7 Key components of the dynamic image analyzer, which is used to measure size, shape, and color in granular material.

these images to determine the shapes and dimensions of the granules, and this analysis is fast enough to process several images per second.[3]

The images are analyzed using the LabView IMAQ software package, which provides the area, perimeter, length, and center position of each of the particles in the image.[4] The scale factor, which is used to convert these dimensions from pixel units to millimeters and square millimeters, is measured beforehand by placing a white disk of known diameter on the inclined surface and measuring its size in pixels. Since the IMAQ software does not provide a built-in function to measure the width of a particle, an algorithm was developed to search for the shortest chord that intersects the center point (Figure 9.8).

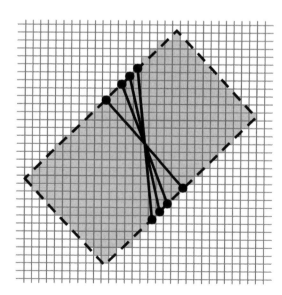

FIGURE 9.8 The width of a granule is measured by finding the shortest chord through the center of its image. The outline of the granule is indicated by the dashed line, and the gray squares depict its digitized representation.

Systematic error in these measurements is primarily caused by variation in illumination, which affects the apparent boundary of the particle.

To test this image-based sensor, a surrogate sample was prepared by chopping Teflon® strands, nominally 1.0 mm in diameter, into over 1,500 individual lengths of 3.0 ± 0.2 mm. Figure 9.9a shows the length and width distributions measured by the imaging system for this surrogate sample. The volume-weighted width distribution is depicted by filled triangles, and the length distribution is depicted by the open triangles; lines have been added in the figure to guide the eye. The observed width (1.0 mm) and length (3.0 mm) have the expected values. When the measurement was repeated, the data (depicted by open circles and filled circles) demonstrated that the reproducibility is excellent; in this case, the deviation in volume fraction was less than 2% in each size class.

Figure 9.9b shows the volume-weighted length distributions for several different batches of granules produced via paste extrusion in a commercial production plant. The length of these granules was changed by monotonically increasing one of the process operating parameters between batches. This parameter has a profound effect on granule

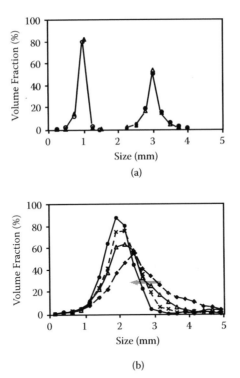

(a)

(b)

FIGURE 9.9 Volume-weighted size distributions measured by the instrument of Figure 9.7: (a) the width (filled triangles) and length (open triangles) distributions of a Teflon® granule reference sample; repeat measurements are depicted by open and filled circles, respectively; (b) the change in length distributions of product granules obtained by increasing one of the process parameters; the four samples are identified in chronological order by diamonds, triangles, crosses, and circles. Adapted from Scott, 2005.

TABLE 9.1   Composition of Test Samples

Sample	Extruded Product		Pan-Granulated Product	
	Amount (g)	Mass Fraction	Amount (g)	Mass Fraction
1	6.00	74.1%	2.10	25.9%
2	3.90	50.6%	3.80	49.4%
3	2.80	31.8%	6.00	69.2%

length, as shown in the figure. Sample uniformity is indicated by the relative width of the length distributions; here the most uniform sample is the one designated by circles.

This sensor has also been able to distinguish between different shapes of individual particles (Scott et al., 2001). To demonstrate this ability, three test samples were prepared by mixing commercial product from a pan granulation process with product from a paste extrusion process. The pan-granulated particles were ellipsoidal pellets roughly a millimeter in diameter with a length of less than about 1.5 mm. The paste-extruded particles were cylindrical, with lengths ranging from 0.5 mm to 5.0 mm and diameters of about 1.0 mm. The mass and mass fractions of the two components are summarized in Table 9.1 for each of the three test mixtures (Sunshine et al., 2005).

The test samples were analyzed by the instrument from Figure 9.7, which calculated the ratio of area to perimeter (A/P) and the ratio of length to width (L/W) for each of the individual granules. When these two ratios are plotted as coordinate pairs, as shown in Figure 9.10, it is observed that the (A/P) ratio tends to be significantly higher for ellipsoids than for cylinders. This result is due to the fact that an ellipse has a more rounded, and therefore shorter, perimeter than a rectangle of the same length and width. By defining a boundary between the two clusters shown in Figure 9.7, it is possible to differentiate between the particles according to their shape. Using the measured width W_i and length

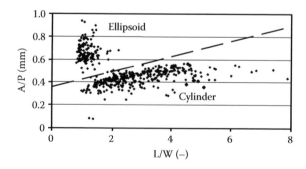

FIGURE 9.10 A plot of the ratio of area to perimeter (A/P) versus the ratio of length to width (L/W) for test sample 1 shown in Table 9.1; the dashed line separates the cylindrical granules from the ellipsoidal granules. From Sunshine et al. (2005).

TABLE 9.2 Comparison of Expected Versus Observed Mass Fractions

	Cylindrical Particles (from extrusion)		Ellipsoidal Particles (from pan granulation)	
Sample	Expected Mass Fraction	Measured Mass Fraction	Expected Mass Fraction	Measured Mass Fraction
1	74.1%	76.7%	25.9%	23.3%
2	50.6%	44.3%	49.4%	55.7%
3	31.8%	29.7%	69.2%	71.3%

L_i of each granule, the total mass of the cylindrical shapes in each sample was estimated from

$$M_C = \rho_C \sum_{i=1}^{N_C} \pi \left(\frac{W_i}{2} \right)^2 L_i \tag{9.6}$$

where N_C is the number of cylinders, and ρ_C is their density (1330 kg·m⁻³). Likewise, the total mass of ellipsoidal shapes in the sample was estimated from

$$M_E = \rho_E \sum_{i=1}^{N_E} \frac{4\pi}{3} \left(\frac{W_i}{2} \right)^2 \left(\frac{L_i}{2} \right) \tag{9.7}$$

where N_E is the number of ellipsoids, and ρ_E is their density (1210 kg·m⁻³).

Finally, the mass fractions were calculated and compared in Table 9.2 with the known values that were shown in Table 9.1. This data is also plotted in Figure 9.11, where the

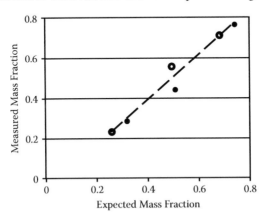

FIGURE 9.11 Mass fraction measured with the instrument shown in Figure 9.7 versus the expected values provided by Table 9.1; open circles denote the data for the ellipsoids produced by pan granulation, and the filled circles denote the data for the cylinders produced by paste extrusion; the dashed line is a linear fit to this data, where $R^2 = 97\%$.

dashed line shows the linear fit to the data; clearly there is good agreement between the expected and measured amounts of the two components. The two data points farthest from the line are both from the measurement of sample 2, which has equal amounts of the two components. At that point the absolute measurement error in mass fraction is about 6%, but the error is only about 3% for the other two samples. Evidently this imaging technique can mistake small cylinders for ellipsoids when the aspect ratio L/W is close to 1, and this slight bias becomes more pronounced when the amounts of cylinders and ellipsoids are nearly equal.

This sensor has also been able to identify granules and other particles by color. The versatility of this approach, which can identify granules by size, shape, and color, has been helpful in understanding the paste extrusion process. The type of data presented here is used for process monitoring at several plant locations.

9.6   Case Study: Media Milling

Agitated media mills are used throughout industry for size reduction and dispersion of a variety of particles, such as pigments, polymers, pharmaceuticals, and agricultural chemicals. Such mills use rotating agitators to stir and thereby fluidize a bed of grinding beads, which typically fill 80% of the volume of the grinding chamber. The nominal size of these beads (called grinding media) ranges from 0.2–3 mm, and the beads may be glass, ceramic, or metallic. The particles to be milled are mixed with liquid to make a slurry that fills the remaining volume in the chamber. Energy is transmitted from the agitator to the slurry and the beads; particles in the slurry break when they are nipped between colliding beads (Kwade, 1999). Some mills pump slurry through the chamber continuously and use a retainer screen to keep the media in the grinding chamber. Batch mills are operated as a closed system, and the media is separated from the product at the end of the process.

Optimum grinding and energy utilization occur when the grinding media are uniformly distributed throughout the mill (Weit, 1987). When all of the beads are separated from each other by liquid, they are able to flow through the chamber as independent particles (although they continue to interact with each other via collisions); beads in this state are said to be *fluidized*. Experience shows that when the flow rate of the slurry through a continuous mill exceeds a critical value, the fluid forces the grinding beads to pack at the retainer screen; this loss of bead fluidization causes screen wear, media wear, an increase in power consumption, and an overall decrease in grinding efficiency. Likewise, if the beads are not fully fluidized in a vertical batch mill, many of them remain on the bottom of the mill and wear grooves into the chamber walls. Grinding under such conditions is ineffective. Successful grinding requires both effective particle capture between the grinding beads and sufficient impact intensity. Bead fluidization influences both capture statistics and collision intensity, and therefore adequate fluidization is a prerequisite for optimum grinding.

Process parameters such as bead size, bead density, bead filling, fluid viscosity and rotational speed affect the level of fluidization, but there is no direct way to measure it using conventional means. An interesting offline application of the ECT technique mentioned in section 9.3 is to measure axial and radial bead distributions of grinding beads and the corresponding fluidization in vertical media mills (Scott & Gutsche, 1999). As demonstrated below, this technique provides a versatile tool to determine the optimum operating conditions for agitated mills.

The vertical mill used in this study is depicted in Figure 9.12; it is constructed entirely of non-conducting materials in order to be compatible with the ECT imaging system. The mill consists of a Plexiglas® tube with an outer diameter of 14.6 cm and an inner diameter of 13.3 cm. The total length of the tube is about 30 cm. A Plexiglas® plate is cemented inside the tube at a distance of 11.4 cm from the bottom. This plate defines the bottom of the milling chamber and provides physical access for the sensing plane of the ECT sensor as shown in the figure. The ECT system used here is a one of the very first commercial units.[5] It has a sensor with an inner diameter of 15.2 cm, allowing it to fit around the vertical mill housing and slide along the axis of the mill. This mill is small by industrial standards but it exhibits the same operational behavior seen in larger mills. As a research tool it provides a platform for quickly measuring the amount of fluidization produced by a variety of operating conditions and agitator designs.

The mill is closed at the top, and the agitator shaft protrudes from a small opening in the lid. The purpose of the agitator is to fluidize the grinding beads and to supply them with the kinetic energy needed to break the particles in the slurry. The shaft of the agitator is connected to a variable speed motor via a torque transducer. This transducer measures torque and rotational speed so that the power input to the mill can be measured. The mill is completely filled to avoid the formation of a vortex. Corn oil is used as a surrogate for the particle slurries, and the grinding media are 1 mm ceramic beads.

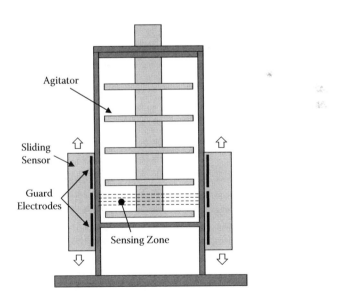

FIGURE 9.12 A cross-section of the vertical mill and the ECT sensor. Adapted from Scott (2005).

Several different agitators have been studied. They are similar to typical designs actually used in industrial mills, but in order to be compatible with the ECT sensor, they are made of Delrin® engineering polymer. Two common designs are the *disk agitator* and *pin agitator*, and representative data obtained with these agitators is shown below. The disk agitator has five removable solid disks, each 12.8 cm in diameter and 0.6 cm thick. The disks are centered on a 3.2 cm diameter shaft, with a gap of 2.5 cm between the disks. The pin agitator has a shaft 2.5 cm in diameter, with six 1.3 cm diameter pins (i.e., rods) installed in mounting holes drilled through the shaft at right angles. The pins are 10 cm long, and they are mounted in an alternating pattern (E–W, N–S, E–W, etc.) along the shaft at a spacing of 2.5 cm.

The tomography system measures the electrical capacitance between every possible pair of electrodes in the array and uses this data to determine the local dielectric constant (as a function of position in the cross-sectional plane) via a tomographic reconstruction. Since the dielectric constants of the fluid (ε_f) and media (ε_m) are fixed, the local bead fraction α can be calculated from

$$\alpha = \frac{(\varepsilon - \varepsilon_f)}{(\varepsilon_m - \varepsilon_f)} \tag{9.8}$$

where the local dielectric constant ε is the result of the tomographic reconstruction at that point in the cross-section. After the system is calibrated against a fully packed media bed (which by definition has a bead fraction of 1), a cross-sectional image showing the bead fraction at every point in that plane can be generated by applying equation 9.8 to every pixel in the original tomogram.

These tomographic images made it possible to determine the radial distribution of the grinding beads. A typical result is shown in Figure 9.13a for the disk agitator rotating at 600 revolutions per minute (rpm). The centripetal force imparted by the agitator tends to push the beads outward, thus increasing the bead fraction near the wall of the mill. This effect is clearly seen in the figure. The ECT images obtained for this agitator show that its design impedes the upward flow of beads and that it never fully fluidizes the grinding media. Much better results were obtained with the pin agitator (Scott & Gutsche, 1999).

The axial distribution is determined by integrating the bead fraction across the sensing plane at a given axial position. Figure 9.13b shows the axial distribution of beads for the pin agitator at a speed of 586 rpm. The bead fraction is approximately constant at 85% over the lower 8 cm of the milling chamber and then decreases to about 50% near the top. This result is due both to poor mixing at the top of the chamber and to gravity, whose downward pull must be overcome by momentum transferred in collisions between the beads. Higher agitation speeds increase the collision rate and send material higher in the vertical mill. In a way, the media behave as atoms of an ideal gas in a gravitational field; beads in the upper levels are supported by collisions with those at the bottom, where the bead fraction is highest. By integrating the data shown in Figure 9.13b for the bead fraction along the axis of the mill, we find the average bead fraction in the mill is about 80%, which is the value expected from conservation of bead volume.

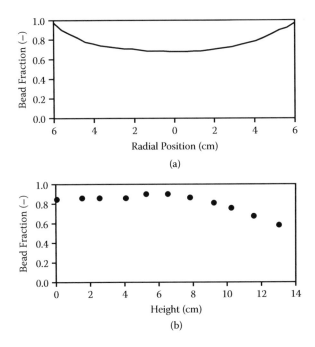

FIGURE 9.13 Results from tomographic imaging of the vertical mill: (a) radial distribution of beads obtained with a disk agitator spinning at 10 s^{-1} (600 rpm); (b) axial distribution of beads obtained with a pin agitator spinning at 9.8 s^{-1} (586 rpm); the bead volume fraction is a function of axial position (height). From Scott & Gutsche (1999).

These results demonstrate that bead fluidization can be monitored by calculating the bead fraction from ECT images. We can define the fluidization parameter f to be a number ranging from 0% to 100%, where 0% corresponds to complete packing of the media) and 100% corresponds to the fully fluidized state. If the overall bead volume is 80%, then the relation between local bead packing fraction α and local fluidization f is given by (Scott & Gutsche, 1999)

$$f = 5(1-\alpha) \tag{9.9}$$

Since local bead fraction, and therefore fluidization, is shown to be a function of position in the mill, we must specify the position at which fluidization is measured. If the average bead fraction $\langle\alpha\rangle$ is obtained by integrating α over the sensing plane at a given height, then the fluidization at that height is

$$f_h = 5(1-\langle\alpha\rangle) \tag{9.10}$$

Figure 9.14 shows the fluidization that is produced by the pin agitator as a function of input power, which is calculated from the torque and speed of the agitator shaft.

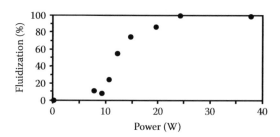

FIGURE 9.14 The observed fluidization of the grinding beads as a function of input power for the pin agitator. From Scott & Gutsche (1999).

The fluidization is measured from equation 9.10 using ECT data obtained at a height of 2.5 cm. It was observed that the mill's motor tends to stall at power inputs below 7 W, and at least 10 watts of power are needed to begin fluidizing the beads (this point occurs at an agitator tip speed of roughly 0.33 m/s). Fluidization increases with power input until about 25 W is reached; further increases in power input do not improve fluidization (which is essentially complete) and only serve to generate more heat in the process. Based on these observations, the conclusion is that the optimal power input in this case is about 25 W. This process imaging technique provides the only direct means of quantifying fluidization of media in mills, and it has clear implications for future process improvements.

Suggested Reading

Castleman, K. (1996). *Digital Image Processing.* Englewood Cliffs NJ: Prentice Hall.

Herman, G.T. (1980). *Image Reconstruction from Projections.* New York: Academic Press.

Rosenfeld, A. and Kak, A. (1982). *Digital Picture Processing.* New York: Academic Press.

Scott, D.M. and McCann, H. Eds. (2005). *Process Imaging for Automatic Control.* Boca Raton, FL, CRC Press.

Scott, D.M. and Williams, R.A. Eds. (1995). *Frontiers in Industrial Process Tomography.* New York: Engineering Foundation.

Williams, R.A. and Beck, M.S. Eds. (1995). *Process Tomography.* Oxford: Butterworth-Heinemann.

10

Thickness Gauging

Reliable online measurements of thickness (also known as *gauge*) are important to the production of many products that are sold in the form of sheets or films. Accurate gauge information is needed not only to control the process, but also for quality control. Variability in the thickness is especially problematic in electronic material applications—such as dielectric films for capacitors, where the changes in capacitance value are directly proportional to changes in the thickness of the film. Likewise, in optical applications the phase shift in transmitted light is determined by film thickness, and variations may cause unwanted visual effects. Gauge is traditionally measured in units where 100 gauge equals 1 mil (0.001 inches), or 25.4 micrometers. Sheet products are usually produced on a semicontinuous production line, and measurements may be made in the *transverse direction* (TD)—that is, across the width of the production line—or in the *machine direction* (MD), as shown in Figure 10.1.

This chapter focuses on the gauging of polymer films using sound and light. The general measurement concepts introduced here can be applied to other gauging applications such as lumber, paper products, or even metal sheets.

10.1   Radiation Gauges

Typical gauging applications rely on sensors that measure the amount of attenuation in a beam of radiation passing through the sample (Figure 10.1). The radiation can be provided by a variety of sources depending on the application. Relatively low-activity (often less than 100 μCi) radioactive sources are used to generate beta, gamma, and X-rays for gauging opaque samples; infrared light can be used for gauging material that is transparent in the infrared region. In passing through the sample, this radiation is attenuated so that the intensity I of the transmitted radiation is given by

$$I = I_0 e^{-\mu t} \tag{10.1}$$

where I_0 is the intensity of the incident radiation, t is the path length through the material (i.e., the film thickness), and μ is a linear attenuation coefficient that generally depends

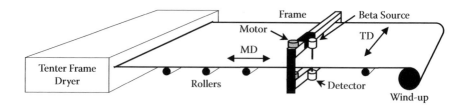

FIGURE 10.1 Film casting operation with a traditional thickness gauge based on attenuation of radiation.

on the density and composition of the film and the energy and type of radiation. For many practical applications, μ is effectively fixed so that

$$t = \left(\frac{1}{\mu}\right) \ln\left(\frac{I_0}{I}\right) \tag{10.2}$$

Sensors of this type measure the transmitted radiation, assume a value for the linear attenuation coefficient, and calculate the thickness t based on equation 10.2. It should be noted that any attenuation due to the column of air between the radiation source and detector is assumed to be negligible; this assumption is incorrect in the case of ultrathin films.

10.2   Ultrasonic Thickness Gauging of Ultrathin Films

Traditional film thickness gauges based on infrared or beta radiation do not work very well with film only a few micrometers thick. At that thickness, there is very little infrared absorption, and the beta gauge (which is sensitive to mass) responds primarily to the air column rather than the film. Therefore, another technique is needed to gauge ultrathin films.

A technique suitable for ultrathin films is based on the measurement of ultrasonic transmission (Lefebvre et al., 1988). The basic concept is illustrated in Figure 10.2 (note the similarity with the radiation gauge shown in Figure 10.1), where a transducer emits

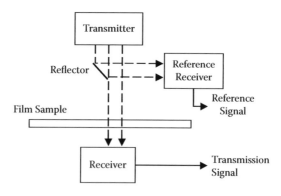

FIGURE 10.2 A thickness gauge based on attenuation of ultrasound.

high frequency sound that passes through the film. This sound has a center frequency of about 215 kHz, so (unlike the megahertz ultrasound used in chapter 8) it can propagate for short distances through air. Two receivers are used to detect the ultrasound: the one below the film records the amount of sound transmitted through the polymer, and the other one monitors a portion of the sound incident on the film. Thin samples transmit the sound readily, but thicker samples tend to reflect rather than transmit the sound. As shown below, the relationship between ultrasonic transmission and sample thickness can be quantified and used as the basis of a thickness gauge. The focus here will be on the design of a practical noncontact ultrasonic device that is used in a production site to determine the thickness of polymer film a few micrometers thick, several meters wide, and moving faster than several meters per second.

10.2.1   Theory of Measurement

As discussed in chapter 3, *ultrasound* is the name given to high-frequency sound waves. The waves are local regions of high and low pressure due to the compression and rarefaction of the air as the disturbance (sound) propagates away from the source. The speed of sound in dry air is about 340 m/s; the wavelength of air-born ultrasound (30 kHz or higher) is therefore roughly a centimeter or less. Since ultrasound is a wave, it is reflected and refracted at surfaces, and it is diffracted upon passage through apertures. When ultrasound impinges on a thin film, some of the ultrasound is transmitted and some is reflected.

Using the arrangement shown in Figure 10.2, one can measure the amount of ultrasound transmitted by the thin film sample. Using the thin plate approximation derived in section 10.4, it is possible to calculate the expected transmission for a thin film of a given thickness. There it is shown that the transmission coefficient T of a thin film is given by

$$T = \sqrt{1 + (\alpha f t)^2} \qquad (10.3)$$

where f is the frequency of the ultrasound, t is the thickness of the sample, $\alpha \equiv \pi \rho_1 / c_0 \rho_0$, ρ_1 is the density of the sample, ρ_0 is the density of air, and c_0 is the sound speed in air. The approximations used to get this result are that the film is very thin compared to the wavelength of sound in the material and that the acoustic impedance (the product ρc) of the film is much greater than that of the surrounding air. Both of these approximations hold for the frequencies and thickness range discussed here. To measure film thickness, we can measure the transmission coefficient and then invert equation 10.3 by using assumed values for f and α. The errors introduced by uncertainties in these values will be discussed in the next section.

Assuming that the film absorbs no ultrasonic energy (i.e., $R^2 + T^2 = 1$), it is easily shown that the reflection coefficient R is given by

$$R = [1 + (\alpha f t)^{-2}]^{-(1/2)} \qquad (10.4)$$

with the same definitions used above. Thus, both the reflection and transmission coefficients are dependent upon frequency, film thickness, and the properties of the film and the surrounding air. It should be noted that the reference detector shown in Figure 10.2 does not measure reflectance; it only monitors the initial ultrasonic signal.

In theory, either transmission or reflection measurements could be used to determine the thickness of the film by solving for the thickness t in either equation 10.3 or 10.4. From equation 10.3 we can deduce that

$$t = \left(\frac{1}{\alpha f}\right)\frac{\sqrt{1-T^2}}{T} \tag{10.5}$$

A similar expression can be found to relate the reflection R to thickness. These expressions show that thickness does not vary proportionally to either T or R. Instead, the thickness is exactly proportional to the ratio $(R/T) = (\alpha f t)$. However, it can be shown that for small values of T (where one can neglect factors of T^2) the thickness is approximately inversely proportional to the transmission coefficient:

$$t = \left(\frac{1}{\alpha f}\right)\left(\frac{1}{T}\right) \text{ for } T \ll 1 \tag{10.6}.$$

The relative error introduced by the approximation in equation 10.6 is less than 1% for $T = 0.1$, and it is about 5% for $T = 0.3$. Assuming a transducer frequency of 215 kHz, a polyethylene terephthalate (PET) film with a thickness of 5 μm would have a transmission coefficient of $T = 0.1$; at a thickness of 1.5 μm the coefficient would be $T = 0.3$. Thus, one would expect that ultrasonic thickness measurements could use the approximation (in equation 10.6) for films thicker than several micrometers; for films as thin as 1.5 μm the error introduced by the approximation would be 5%, so it would become necessary to use the exact solution given by equation 10.5.

The relative error introduced by the approximation in equation 10.6 is shown in Figure 10.3 as a function of film thickness (in micrometers). The relative error is defined

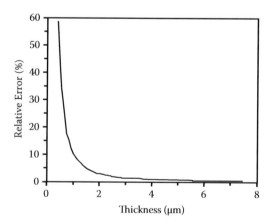

FIGURE 10.3 Relative error associated with the approximation used in equation 10.6.

as the difference between the approximation in equation 10.6 and the exact solution in equation 10.5, divided by the exact solution. For purposes of calculation, the following values have been assumed:

$$\rho_1 \text{ (density of the film)} = 1300 \text{ kg/m}^3$$

$$\rho_0 \text{ (density of air)} = 1.21 \text{ kg/m}^3$$

$$c_0 \text{ (sound speed in air)} = 345 \text{ m/s}$$

$$\alpha \equiv \pi \rho_1 / c_0 \rho_0 = 9.86 \text{ s/m.}$$

It is evident in Figure 10.3 that the approximation breaks down rapidly for films thinner than 2 μm.

10.2.2 Measurement Sensitivity

The sensitivity of the gauge can be defined as the ratio between the relative change in transmission ($\Delta T/T$) to a given relative change in thickness ($\Delta t/t$). Differentiating equation 10.3 with respect to thickness, one obtains

$$\frac{dT}{dt} = -[1 + (\alpha ft)^2]^{-(3/2)} (\alpha ft)^2 \left(\frac{1}{t}\right) \tag{10.7}$$

After substitution of equation 10.3 into equation 10.7, a simple rearrangement yields the sensitivity formula:

$$\left|\frac{\Delta T/T}{\Delta t/t}\right| = \frac{(\alpha ft)^2}{1 + (\alpha ft)^2} \tag{10.8}$$

For a 5-micrometer film, transducers operating at 215 kHz should yield a sensitivity of over 99%. Thus, an excellent correspondence is expected between changes in film thickness and the resulting changes in the transmission of ultrasound. Figure 10.4 shows the transmission sensitivity as a function of film thickness, assuming a transducer frequency of 215 kHz.

It is evident from Figure 10.4 that below a thickness of 1 micron the sensitivity drops sharply, suggesting that a higher frequency will be required to measure thinner films. It should be noted in passing that the sensitivity for reflection is the complement to that for transmission; thus, the sensitivity for reflection measurements increases as the thickness decreases. Reflection measurements are recommended for gauging films that are less than 1 μm thick.

10.2.3 Effect of Ambient Conditions

Since the transmission of ultrasound is dependent upon the local acoustic conditions, changes in air temperature or pressure (which affect the local density of the air) are

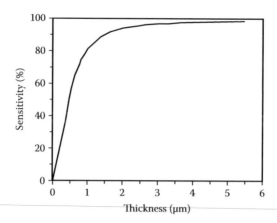

FIGURE 10.4 Measurement sensitivity versus film thickness (from equation 10.8).

expected to have an impact on the recorded signal. In order to get accurate readings, fluctuations in the ambient conditions must be kept to a minimum so that the transmission signal is more responsive to thickness than to ambient factors.

To determine the sensitivity of transmission measurements to changes in air temperature and pressure, one takes the differential of equation 10.3, holding thickness fixed. The definition of α is noted to include the density of air ρ_0, the sound speed in air c_0, and the density of the sample ρ_1 (which is treated as a constant). After dividing this differential by equation 10.3, we find that

$$\frac{\Delta T}{T} = \frac{(\alpha ft)^2}{[1+(\alpha ft)^2]}\left(\frac{\Delta\rho_0}{\rho_0} + \frac{\Delta c_0}{c_0}\right) \tag{10.9}$$

Equation 10.9 represents the relative change in the transmission coefficient due to the combined relative changes in air density and sound speed, which are caused by ambient effects. Comparing equation 10.9 to equation 10.8, it is evident that the transmission coefficient is just as sensitive to the combined variations in density and sound velocity as it is to changes in film thickness. The change in apparent thickness t due to the change in transmission coefficient T can be calculated by taking the differential of equation 10.5 with everything but T constant. Dividing the differential by equation 10.5, one finds that

$$\frac{\Delta t}{t} = -\left[\frac{1+(\alpha ft)^2}{(\alpha ft)^2}\right]\left(\frac{\Delta T}{T}\right) \tag{10.10}$$

After combining equation 10.9 with equation 10.10, it is seen that the relative error in the apparent thickness is given by

$$\frac{\Delta t}{t} = -\left(\frac{\Delta\rho_0}{\rho_0} + \frac{\Delta c_0}{c_0}\right) \tag{10.11}$$

The relative error in the thickness measurement is thus equal to the combined relative variations of density and sound velocity. Note that ρ_0 and c_0 should be treated as intermediate variables, because they depend upon temperature and pressure. The next step is to examine the temperature and pressure dependence of these variables.

Temperature Dependence

Treating air as an ideal gas, the speed of sound is given (Currie, 1974, p. 330) as

$$c_0 = \sqrt{\gamma RK} \tag{10.12}$$

where R is the universal gas constant, γ is the ratio of the specific heat capacity at constant pressure to that at constant volume, and K is the temperature (so designated to distinguish it from the transmission coefficient T). Taking the differential of equation 10.12 and dividing the result by equation 10.12 yields

$$\frac{\Delta c_0}{c_0} = \frac{1}{2}\left(\frac{\Delta K}{K}\right) \tag{10.13}$$

Starting with the ideal gas law $P = \rho RK/M$, where M denotes the molecular mass, it is likewise easy to show that

$$\frac{\Delta \rho_0}{\rho_0} = -\left(\frac{\Delta K}{K}\right) \tag{10.14}$$

It can be shown from equations 10.11, 10.13, and 10.14 that temperature variations introduce a relative error in the thickness measurement given by

$$\frac{\Delta t}{t} = \frac{1}{2}\left(\frac{\Delta K}{K}\right) \tag{10.15}$$

Thus, a 1°C temperature shift in the vicinity of the transducers will result in a 0.2% increase in the apparent film thickness. Achieving measurement accuracy of 1% will require that the temperature in the sound path is known to within 5°C.

Pressure Dependence

The barometric pressure also affects the measured thickness. Equation 10.12 predicts that the sound velocity in an ideal gas is governed by the temperature only. Thus, any change in pressure will not change the value of c_0:

$$\frac{\Delta c_0}{c_0} = 0 \tag{10.16}$$

Differentiating the ideal gas law $P = \rho RK/M$, holding temperature K constant, one can show that the relative change in density equals the relative change in pressure:

$$\frac{\Delta \rho_0}{\rho_0} = \frac{\Delta P}{P} \tag{10.17}$$

By combining equations 10.11, 10.16, and 10.17, one finds that pressure variations introduce a relative error in the thickness measurement as given by

$$\frac{\Delta t}{t} = -\frac{\Delta P}{P} \tag{10.18}$$

Barometric pressure can change by as much as 10%, causing a large long-term error in the measured thickness. In order to make accurate thickness measurements, it is necessary first to measure the barometric pressure and to make the appropriate correction to the value of ρ_0 used in the calculations. With a capacitance manometer to measure pressure automatically, it should be possible to reduce this source of measurement error to a small fraction of 1%.

Humidity Dependence

Finally, one can consider the effect of humidity on these measurements. From available data on the velocity of sound at 80 kHz (Lide, 1991, chap. 14, p. 36ff), one notes that c_0 increases only 0.3% as the relative humidity is increased from 0% to 100% (at room temperature). Therefore the effect of humidity on c_0 is negligible under typical atmospheric conditions. The effect of humidity on the density of the air is determined by considering the partial pressure of the water vapor present. The amount of water vapor absorbed by the air is limited by the vapor pressure of water at that temperature. At 100% relative humidity, the partial pressure of the water vapor is equal to the vapor pressure of water. At a temperature of 30°C, the vapor pressure of water is 4.25 kPa (Lide, 1991, chap. 6, p. 9). By application of the ideal gas law, one finds the corresponding density of the water vapor to be

$$\rho = \frac{PM}{RK} = \frac{(4250)(18)(10^{-6})}{(8.314)(303)} = 0.030 \text{ kg/m}^3 \tag{10.19}$$

for a relative humidity of 100%. It should be noted that this value is 2.5% of the density of dry air (at standard conditions), and under high humidity conditions there is a fairly small decrease in density (over dry air). Even at 40°C, saturated air is only 7% less dense than dry air. From equation 10.11 it is evident that humidity will affect thickness measurements primarily through the decreased density of the air. The actual change in apparent thickness will depend upon the relative humidity and the temperature, as shown in Figure 10.5.

Note that a decrease in density leads to an increase in apparent thickness. Each contour line in Figure 10.5 represents an additional 1% change in the apparent thickness

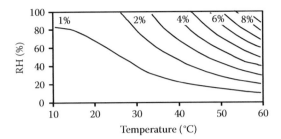

FIGURE 10.5 Contour plot depicting the percentage increase in apparent thickness as a function of both temperature (in °C) and relative humidity (in %).

due to the decrease in air density from humidity. The effect of temperature on the density of dry air, shown by equation 10.14, has not been included in this calculation.

The results of this analysis suggest that of the three environmental factors—temperature, pressure, and relative humidity—the most significant source of measurement error is the barometric pressure. Temperature and relative humidity will contribute additional relative errors of less than 2% each (under normal atmospheric conditions). It is not possible to achieve accuracy in the measurement of thickness less than 5% under ambient conditions without taking these fluctuations into account. Monitoring the temperature and barometric pressure near the transducers and correcting the values of ρ_0 and c_0 used in the calculation of the thickness can improve the accuracy. Monitoring or maintaining the humidity level within the measurement area should also reduce the error by approximately 1%.

It should be stressed that all of these considerations assume a uniform environment between the transducers, so any measurement of ambient conditions should accurately reflect the local temperature and pressure of the air in the path of the ultrasound. This assumption is not necessarily valid in a manufacturing setting. Local disturbances, such as blasts of hot air emanating from the line, profoundly affect the propagation of ultrasound. Thus, the local environment between the transducers must be protected as well as monitored. The effects of ambient conditions on ultrasonic measurements have been seen at the plant: the sensor, although relatively stable in operation in the lab, exhibited output fluctuations when initially mounted online. These fluctuations were found to be due to hot air currents generated by the line equipment. Subsequent installation of an isolation chamber around the gauge significantly reduced the observed variation in output.

Even the addition of an isolation chamber may not completely protect the gauge from spurious effects. The moving film carries along with it a thin boundary layer of warm air from the process. When this boundary layer enters the measurement chamber, which is likely to be at a different temperature, the boundary layer itself will reflect a small fraction of the ultrasound. The magnitude of this effect depends on line speed and temperature differential as well as atmospheric conditions. A proper treatment of this possibility is outside the scope of the present discussion, since the boundary layer is difficult to characterize without specific information about the placement of apertures,

heat sources, and equipment throughout the line. Although the thin-plate approximation given in section 10.4 does not take thermal layers into account, an extension of the
theory could easily be made. For the present, the measurement errors introduced by the
boundary layer are ignored.

10.2.4   Transducer Effects

The theory developed so far has assumed that the ultrasonic waves are perfect plane
waves of continuous duration at a single frequency. In practice, the frequency content
and spatial extent of the acoustic field may differ from the ideal. The shape of the acoustic field generated by the physical transducer has important ramifications for the operation of the ultrasonic gauge.

Piezoelectric transducers generate ultrasound when an electric field is created across
their piezoelectric crystal (which is generally cylindrical). The field causes a distortion
of the crystal lattice, thereby displacing the face of the crystal a distance proportional to
the voltage applied to the transducer. A high frequency signal generates an oscillating
electric field, causing the crystal face to move back and forth as a piston at the driving
frequency. When this motion is coupled to the air (or some other medium), the resulting
longitudinal waves become ultrasound. This so-called piston source has a distinctive
radiation pattern that is determined by the diameter of the crystal and the wavelength
of the ultrasound.

Radiation Pattern

In the far field, the radiation pattern of a piston source looks like the diffraction pattern
of an infinite plane wave that has passed through a circular aperture (Krautkramer &
Krautkramer, 1983, p. 62 ff). The directional characteristics are given (see Olson, 1947,
p. 38) as

$$P(\theta) = \frac{2J_1(x)}{x} \quad \text{with } x \equiv \frac{\pi D}{\lambda} \sin\theta \qquad (10.20)$$

where $P(\theta)$ is the acoustic pressure observed at angle θ, normalized to the pressure at
$\theta = 0°$; θ is the angle of observation, measured with respect to the normal; J_1 is the
Bessel function of the first order; λ is the wavelength; and D is the diameter of the piston
(i.e., the transducer crystal). The radiation pattern predicted by this equation is shown
in Figure 10.6 for several values of the parameter D/λ.

Several observations can be made from Figure 10.6. For diameters smaller than the
wavelength, the transducer gives a wide-angle response. In particular, for diameters less
than a quarter wavelength the transducer response is nearly isotropic. For transducers
with a diameter equal to the wavelength, the response albeit broad begins to become
more directional, forming a well-defined lobe in the forward direction. For $D/\lambda = 2$, the
main lobe becomes more directional, and two side lobes develop. Finally, transducers
with diameters much greater than the wavelength emit ultrasound in highly directional
patterns. The angular width of these lobes decreases as the diameter is increased (relative to the wavelength). Additional side lobes develop, greatly complicating the off-axis

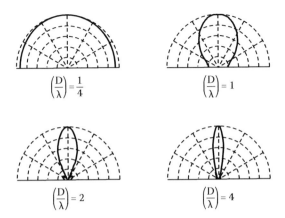

$$\left(\frac{D}{\lambda}\right)=\frac{1}{4}$$

$$\left(\frac{D}{\lambda}\right)=1$$

$$\left(\frac{D}{\lambda}\right)=2$$

$$\left(\frac{D}{\lambda}\right)=4$$

FIGURE 10.6 Transducer radiation patterns given by equation 10.20 at various values of D/λ.

emission pattern as the diameter is increased. (Although the scale of Figure 10.6 is too gross to show it, there are eight side lobes in the last plot.)

In order to determine the angular extent of the main lobe, note that the first zero of the Bessel function in equation 10.20 occurs at $x = 3.83$. Thus, the angle θ_0 at which this zero occurs is given (see Hueter & Bolt, 1955, p. 65) as

$$\theta_0 = \sin^{-1}\left(1.22\frac{\lambda}{D}\right) \tag{10.21}$$

Note that this angle θ_0 is really just the half-angle of the ultrasonic beam. Thus, the size of the beam is determined by the frequency and the size of the transducer. For narrow-beam transducers, one can use the small-angle approximation to find the diameter S of the beam at a distance Z from the source:

$$S \approx (2.44)\frac{Z\lambda}{D} \tag{10.22}$$

This spot size (in equation 10.22) is the one usually given in the literature, but it overestimates the size of the area actually interrogated in the measurement of film thickness. The ultrasonic gauge reading is an average thickness over an effective spot size. In order to increase the spatial resolution of the gauge, it is necessary to reduce the effective spot size S_e (discussed later in this section). Attempts to limit the size of the beam by placing an aperture in its path would only diffract the beam into larger angles according to equation 10.20.

As a concrete example, consider a particular piezoelectric transducer: the model E-188/215, manufactured by Massa Products of Hingham, Massachusetts. This narrowband transducer operates at 215 kHz and has a total beam angle (measured at the −3 dB point) of 10°. This beam angle is not quite the same as θ_0 since it is measured with

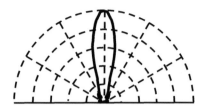

FIGURE 10.7 The radiation pattern calculated for E-188/215 transducers.

respect to the −3 dB point and not with respect to zero pressure. These points occur when $P(\theta) = 0.714$, so that upon substitution into equation 10.20, the vendor's specification implies that

$$(0.714)x = 2J_1(x) \text{ where } x = (0.545)\frac{D}{\lambda} \tag{10.23}$$

Solving this equation graphically one finds that $x = 1.595$, which implies that $D/\lambda = 2.93$ for this transducer. The radiation pattern corresponding to $D/\lambda = 2.93$ is plotted in Figure 10.7. Equation 10.21 shows that the half-angle of the lobe is 24.6°, giving a beam angle of nearly 50°.

Note that equations 10.20–10.23 are based on an analysis of the acoustic emission in the far field. It is well known that for circular pistons with $D/\lambda \gg 1$, the transition from the near field to the far field occurs at a distance $Z_0 \approx D^2/4\lambda$ from the source (Hueter & Bolt, 1955, p. 65). In the present case, $D/\lambda = 2.93$, so that the transition from near field to far field occurs at $Z_0 \approx (0.73)D$. The distances involved in the manufacturing application are well beyond the transition point; therefore the far field analysis is justified.

X-Y Sensitivity

The radiation pattern shown in Figure 10.7 can be used to examine the effects of transducer alignment. Since the determination of thickness is based on a transmission measurement, the alignment of the source and detector is an important consideration. The same type of transducer used to generate the ultrasound can be used to detect the transmitted ultrasound as well. According to the concept of transducer reciprocity, the radiation sensitivity pattern of the transducer acting as a receiver is assumed to be the same as the emission pattern of the transmitter.

As long as the transducers are in perfect alignment, the receiver sees the maximum pressure exerted along $\theta = 0°$ by the transmitter, and the ultrasound is received at the receiver's optimum angle ($\theta = 0°$). If the transducers are not aligned, then the lateral offset (the x-y displacement) introduces a common angle θ between the transducers. The signal strength of the received ultrasound (neglecting the film) will be the product of the transmission strength of the transmitter with the antenna gain of the receiver. These factors are given by the corresponding points on the radiation patterns, as shown in Figure 10.8.

As seen in the figure, if two otherwise aligned transducers suffer a relative displacement Y at a distance Z, then the angle θ of transmission is given by $\tan \theta = Y/Z$. Note that the angle of reception is also θ. Assuming that the same type of transducer is used for both transmitter and receiver, the gain product G (the relative signal strength in the absence of film) as a function of lateral displacement Y and transducer separation Z is given by

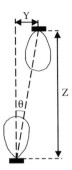

FIGURE 10.8 Stylized radiation patterns showing the effect of transducer misalignment.

$$G(Y/Z) = \frac{4[J_1(x)]^2}{x^2} \quad \text{where } x = \frac{\pi D}{\lambda}\left[\frac{(Y/Z)}{\sqrt{(Y/Z)^2 + 1}}\right] \tag{10.24}$$

Thus, the characteristic length scale in the determination of G is set by the transducer separation Z. This treatment of course assumes that the two transducers are not tilted with respect to each other.

Given the results of equation 10.24, one can determine the effects of transducer misalignment by assuming $D/\lambda = 2.93$ (as derived above) and estimating a separation of $Z = 8$ cm. With these assumptions, the predicted deviation of $G(Y/Z)$ from unity is shown in Figure 10.9, where the coupling loss is simply $1 - G(Y/Z)$.

The significance of Figure 10.9 is not only that a lateral misalignment of a few millimeters substantially lowers the ultrasonic coupling, but also that small changes in the lateral displacement result in large variations in the ultrasonic coupling. In a practical sensor used to gauge film in the transverse direction, the transmitting and receiving transducers are mounted on separate carriages that move along fixed rails. The mechanical linkages coupling the two carriages can introduce a variable displacement between the two transducers as they move across the width of the film. If the nominal alignment is perfect (i.e., the average displacement is zero), then a maximum displacement of 1 mm

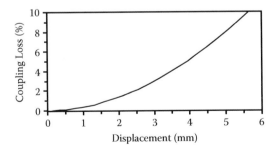

FIGURE 10.9 Coupling loss as a function of transducer displacement Y for $Z = 8$ cm.

introduced by the motion of the carriage will cause only a 0.3% coupling loss. However, if there is a misalignment such that the average displacement is 3 mm (giving a coupling loss of 3%), then an additional 1 mm displacement caused by the carriage motion will cause the coupling loss to increase an additional 2%. Note that a 2% increase in coupling loss yields approximately a 2% increase in apparent film thickness. Therefore, a 1 mm relative displacement between the transducers yields a 0.3% change in apparent thickness when the system is properly aligned, yet the same 1 mm relative displacement yields a 2% change in apparent thickness when the alignment is off by only 3 mm. This error increases as the static alignment worsens. Both the static alignment and the displacement variability (the backlash in the traversing mechanism) thus significantly impact the measurement accuracy.

It is instructive to consider the significance of transducer alignment on the passage of ultrasound through the film. If there is a displacement Y between the transducers, then the sound beam effectively makes an angle θ with the film, as shown in Figure 10.8. Of course, if the emitting transducer is not normal to the film, then the angle of the sound beam also affects the angle of incidence θ. This nonnormal incidence angle affects the transmission and reflection coefficients for the ultrasound. A theoretical treatment of the transmission coefficient as a function of θ has been worked out by Brekhovskikh (1980, pp. 76–80), but the results are too involved to be included here. For the present discussion it is sufficient to show the experimental results obtained by Lefebvre et al. (1988).

In Figure 10.10, the abscissa is the tilt or incidence angle θ and the ordinate is the normalized transmission coefficient given by $T(\theta)/T(0)$. Experimental observations are indicated by squares, with a solid line giving the theoretical prediction. For comparison, the gain product G from equation 10.24 is calculated for $D/\lambda = 2.93$ as a function of θ and included in the figure as a dashed line. It is evident that beam tilt has an impact on the transmission coefficient; the transmission coefficient increases with θ, ultimately increasing by 10% at 30°. Theory predicts a 15% increase in transmission at 30°, but it is extremely unlikely that such large angles would be found in practice. In contrast, the drop in transducer coupling (shown by the dashed line in Figure 10.10) due to misalignment is

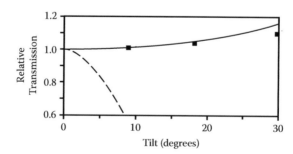

FIGURE 10.10 Normalized transmission coefficient as a function of incidence angle of the beam. Squares depict experimental data from Lefebvre et al. (1988); the solid line is the theoretical prediction from Brekhovskikh (1980), and the dashed line is the gain factor G from equation 10.24 for $D/\lambda = 2.93$, shown here as a function of angle.

much more severe: the signal drops by half at an angle of only 10°. In effect, the misalignment causes the beam to miss the receiver.

The static alignment and dynamic displacement variability of the transducers thus introduce errors in the thickness measurement via a position-dependent radiation coupling loss, resulting in so-called *x-y* sensitivity. In order to reduce the measurement system's sensitivity to *x-y* displacement, it is necessary to change the radiation pattern by replacing the transducers. Referring to equation 10.24 and Figure 10.8, the gain product *G* at a given (*Y/Z*) will be largest for radiation patterns with wide lobes. Therefore *x-y* sensitivity can be reduced by replacing the current transducers with ones that have a large beam angle. Obviously, if a larger beam angle is used, the effective spot size of the area being measured will be larger. For this reason, one might be tempted to use a narrow-beam source in conjunction with a wide-beam receiver to maintain a small spot size. It should be noted that this arrangement would still suffer from *x-y* sensitivity, since the gain product *G* will be dominated by the directional characteristics of the narrow beam.

Effective Spot Size

Returning to the discussion of the effective spot size S_e of the ultrasonic beam, a rigorous consideration of the acoustic coupling shows that in the far field the effective size of the beam at a point midway between source and detector is one half the diameter of the detector, regardless of the separation between the transducers. As shown in Figure 10.11, a film introduced between the source and detector transducers redirects some of the acoustical energy onto the detector. The waves emanating from (or through) the film can be described using Huygen's Principle: considering each point on the film as a source of spherical waves (each with a distinct phase) one builds up the original ultrasonic beam passing through the film. Conceptually, invoking Huygen's Principle gives us a mechanism for scattering the acoustic energy toward the detector, since the spherical wavelets propagate in all directions.

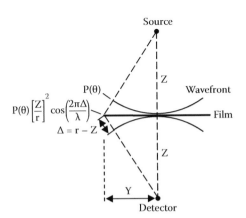

FIGURE 10.11 The phase and amplitude of Huygen's waves originating at the film.

It should also be noted that in the thin-plate approximation, the spherical wavefronts generated by the source pass through the film without losing their spherical character. Thus, there is no spherical aberration (to use the optical analogy), and the relative phase along the wavefront is preserved.

Figure 10.11 shows a spherical wave from the source passing through the film, which is placed midway between source and detector. At an angle θ with respect to the normal, the relative ultrasonic pressure $P(\theta)$ at the wavefront is given by equation 10.20; at the point indicated (on the film), the pressure wave phase is retarded by some amount δ due to the increase in path length. In addition, the Inverse Square Law will reduce the relative pressure amplitude by a factor $(r/Z)^2$, where r is the distance from that point to the source and Z is the source-film distance. Thus, at a given point on the film, the Huygen's pressure wave has an amplitude of $P(\theta)(Z/r)^2\cos(\delta)$. This wave will propagate in all directions and will enter the detector at an angle θ with a detection efficiency $P(\theta)$. The amplitude of the wave at the detector will be reduced further by the Inverse Square Law. It is evident that the phase (relative to the transmitted wave at $\theta = 0°$) of the Huygen's wavelets entering the detector will be retarded by an additional amount; from symmetry it is clear that the additional retardation is also δ. Therefore each point on the surface of the film generates a corresponding wavelet at the detector, where the wavelet has amplitude and phase given by $[P(\theta)]^2(Z/r)^4\cos(2\delta)$. The total received ultrasonic signal can be calculated by integrating all of these contributions over the area of the film.

The task at hand is to find the effective area of film measured by the ultrasonic gauge. Noting that $[P(\theta)]^2$ is simply the gain product G, Figure 10.9 shows that the contributions given by $[P(\theta)]^2(Z/r)^4\cos(2\delta)$ will decrease rapidly off the beam axis. Most of the received signal is therefore due to ultrasound near the beam axis. It was noted earlier that the beam diameter given by equation 10.22 overestimates the area actually measured by the ultrasonic beam. Equation 10.22 gives the diameter of the area exposed to the beam, but the position dependent sensitivity (the gain factor) and the phase of the wavelets are not taken into account. A useful measure of the effective spot size diameter S_e can be made by integrating wavelets over the film area and determining the equivalent size of a disk of unit thickness that includes the same volume. A simple illustration of this concept is shown in Figure 10.12.

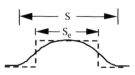

FIGURE 10.12 Spot size versus effective spot size.

In the figure, the solid line represents the amplitude $[P(\theta)]^2(Z/r)^4$ of the wavelet contributions as a function of distance from the beam center. The dashed line is drawn so that its amplitude is the same as the maximum amplitude of the solid line (the wavelet contribution), and the width of the dashed line is chosen so that the area is the same under both curves. This width is defined to be S_e. In the case of film gauging, one can integrate over an area (the plane of the film) in order to find the diameter S_e. The maximum value of the wavelet contribution is unity. Thus, S_e is the diameter of a disk of unit height with a volume equivalent to that under the surface defined by $[P(\theta)]^2(Z/r)^4\cos(2\delta)$. Since $[P(\theta)]^2$ is simply the gain product $G(Y/Z)$ given in equation 10.24,

$$\frac{\pi(S_e)^2}{4} = \int_0^{Z \cdot \tan\theta_0} G(Y/Z)\frac{Z^4 \cos(2\delta)}{(Y^2+Z^2)^2} \cdot YdY \cdot \int_0^{2\pi} d\varphi \qquad (10.25)$$

where the distance r is expressed in terms of Y and Z, and δ is defined by

$$\delta = 2\pi\left(\frac{\sqrt{Y^2+Z^2}-Z}{\lambda}\right) = 2\pi\left(\frac{Z}{\lambda}\right)\left(\sqrt{\left(\frac{Y}{Z}\right)^2+1}-1\right) \qquad (10.26)$$

Picking the separation Z so that it is an integral number m of wavelengths (let $Z = m\lambda$), the phase shown in equation 10.26 can be simplified to

$$\delta = 2\pi m\sqrt{\left(\frac{Y}{Z}\right)^2+1} \qquad (10.27)$$

In equation 10.25 the integration over Y is from zero to the value of Y at the edge of the lobe, which is Z times the tangent of the beam half-angle θ_0 defined in equation 10.21. The integration over the polar angle contributes an additional factor of Y. Simplifying equation 10.25, one finds that

$$(S_e)^2 = 8Z^2\int_0^{\tan\theta_0}\frac{G(\zeta)\cdot\zeta\cdot\cos(2\delta)}{(\zeta^2+1)^2}d\zeta \qquad (10.28)$$

where a change of variables $(Y/Z) = \zeta$ has been made. Equation 10.28 can be solved for S_e by noting the value of θ_0 and substituting equation 10.24 into the integrand.

In the present derivation, estimated values of $D/\lambda = 2.93$ and $\theta_0 = 24.6°$ are used. By numerical integration, the integral in equation 10.28 is found to vary inversely as the square of the parameter m appearing in equation 10.27. For $m > 30$, the value of the integral is approximately $0.285/m^2$. This approximation is valid for the present case, where Z is assumed to be about 8 cm and λ is 1.5 mm. Since by prior definition $Z = m\lambda$, factors of m^2 cancel, and the effective spot size is independent of Z:

$$(S_e)^2 = 8Z^2\frac{(0.285)}{m^2} = (2.28)\lambda^2 \qquad (10.29)$$

$$\therefore S_e = (1.51)\lambda \qquad (10.30)$$

Finally, we recognize that $(D/\lambda) = 2.93$ for these transducers; substituting this ratio into equation 10.30 yields the effective spot size in terms of the transducer (detector) diameter:

$$S_e = (0.515)D \qquad (10.31)$$

Thus, S_e for a film placed midway between the transducers is independent of spacing and is about one-half the transducer diameter.

Although the preceding analysis is based on a particular D/λ ratio, the result of equation 10.31 is found to be generally applicable. Numerical integration of equation 10.28 with other values of D/λ indicates that for $D/\lambda > 2$, the integral is approximately $D^2/32m^2\lambda^2$ or $D^2/32Z^2$. Multiplying by $8Z^2$ as in equation 10.29 and taking the square root, one finds that $S_e \approx (0.5)D$ regardless of separation or D/λ. Referring back to Figure 10.6, it is evident that the width of the ultrasonic beam is determined by the ratio D/λ. Therefore this result is somewhat counterintuitive, since it suggests that the measurement area on a film is no smaller for narrow beams than it is for wide beams.

Since x-y sensitivity is enhanced for transducers with narrow beams, the real significance of this result is that one can use transducers with wider beams (thereby reducing the x-y sensitivity) without sacrificing the spatial resolution of the measurement S_e. Obviously, one cannot make the transducer diameter D arbitrarily small. From equation 10.21 it is seen that the beam half-angle θ_0 goes to 90° for $D/\lambda = 1.22$; therefore, the upper limit of integration in equation 10.28 goes to infinity at this value. For values of D/λ in the range of 1.22 to 2.0, the approximation implicit in equation 10.31 begins to break down; at $D/\lambda = 1.5$, $S_e = (0.563)D$ and at $D/\lambda = 1.3$, $S_e = (0.582)D$. At $D/\lambda = 1.22$, the transducer becomes an omni-directional source, so there is no longer a well-defined beam. In any case, given the transducer diameter D, one can choose the frequency so that the ratio of diameter to wavelength is greater than 2. Thus, measuring the thickness over a small area requires the use of higher frequencies, not smaller transducers.

A word of caution is warranted regarding the interpretation of S_e. It must be remembered that S_e is defined in Figure 10.12 to be the diameter of a disk of unit thickness with a volume equivalent to that under the surface integrated in equation 10.25. Thus, some contributions to the measurement signal are expected to come from an area outside that defined by S_e. For example, equation 10.31 suggests that the ultrasonic sensor has an effective spot size of 2.4 mm, but in fact the thickness measurement will be a convolution of the thickness over a much larger area. This convolution will be weighted to favor the area defined by S_e, but significant thickness variations outside that will still be seen by the transducers (with reduced sensitivity). Thickness variations that are completely outside the diameter S defined in equation 10.22 will not be included in the convolution.

Frequency Response

A final issue concerning the transducers is their frequency response. The foregoing derivations have assumed a single-frequency, continuous ultrasonic wave. In practice, the transducers are operated in tone-burst mode in order to avoid problems with acoustic reverberations. The tone burst must be long enough to excite the transducer to full strength, yet shorter than the transit time of the echoes. It has been observed experimentally that narrow-band transducers such as those used in this application require 15 cycles or more at the appropriate frequency in order to reach maximum output. Allowing a 15 cycle ring-down period, the full tone burst would consist of at least 30 cycles. At a frequency of 215 kHz, this tone burst would last about 140 μs. Since the speed of sound in dry air is 340 m/s, the length of the ultrasonic wave train is nearly 5 cm. If the receiving transducer is 4 cm (for example) from the film, then the round trip distance traveled by an echo is 8 cm. This distance corresponds to 235 μs, or approximately 50 cycles at 215 kHz. Longer duration tone bursts would generate multiple echoes, and these echoes must be avoided. Therefore, the tone burst must consist of (at least) 30 to (at most) 50 cycles at the fundamental frequency of 215 kHz.

The frequency content of these tone bursts is determined not only by the carrier frequency, but also by the length of the burst and the frequency response of the transducer. The frequency response of the receiving transducer also affects the measured response. Strictly speaking, the ultrasonic wave passing through the film is not a plane wave at a well-defined frequency. However, the results of the measurement theory derived in section 10.4 can still be used in the general case by resolving the tone-burst actually used into its spectral components (each at single frequency) and applying the gauging equations 10.5 or 10.6 to the transmission coefficient at each frequency. This approach can be extended to a wide-band measurement wherein an ultrasonic pulse of short duration but wide spectral content is used to measure film thickness. The main advantages of a pulse system are its speed and its ability to exclude echoes. There does not however appear to be any advantage connected with the ability to measure at several frequencies simultaneously: As stated earlier, the ultrasonic method measures the quantity αft, so that measurements of T at various frequencies f cannot be used to remove the dependence of thickness on α.

10.2.5 Summary

This analysis provides a fundamental description of the physical process involved in measuring film thickness with ultrasound, with the emphasis placed on those factors that corrupt the gauge measurements. The dependence of these measurements on both ambient conditions and transducer alignment is striking. Of the three environmental factors—temperature, pressure, and relative humidity—the most significant source of measurement error is barometric pressure. Significant improvements in accuracy can be achieved by measuring the temperature and barometric pressure near the transducers and correcting the density and sound speed values used in the calculation of the thickness.

A potentially more severe source of measurement error is due to transducer misalignment. This misalignment can be either static or dynamic (as in a TD gauge traversal mechanism). Even small (1 mm) amounts of backlash in the mechanical linkages can introduce errors of several percent. Using the formalism developed here, it is possible to estimate the actual amount of error for a given amount of backlash. Thus, the specifications for the mechanical system can be determined once the accuracy required of the gauge measurements is established.

The spot size of the measurement is determined by the size of the transducers rather than the size of the ultrasonic beam. At a given frequency, decreasing the transducer diameter will increase the beam diameter (lobe size), so one might expect the spot size to increase as well. The fact that it does not is a striking result of the present analysis. It should be noted that large beams are compatible with reducing the x-y sensitivity, so there is no trade-off between small spot size and insensitivity to transducer misalignment.

For transmission measurements on films under 2 micrometers thick, the exact solution (equation 10.5) must be used instead of the approximate solution (equation 10.6). Films under 1 micrometer in thickness may best be measured with a reflection type of technique, which allows more sensitivity in that thickness range.

10.3 Optical Thickness Measurements for Single and Multilayer Films

Coextruded laminates and other multilayer films consist of two or more distinct layers of various materials. This type of product is used in packaging, optical displays, and multilayer circuit boards. An example is the three-ply structure shown in Figure 10.13, which depicts a product for use in the electronics industry. The middle layer (about 25 micrometers thick) is Kapton®, while the outer layers are made of adhesive (about 12.5 micrometers thick). This coextruded laminate is used in conjunction with copper foil to construct flexible printed circuit boards. The thicknesses of the layers are determined by the extruder die settings, and it is clearly advantageous to be able to adjust the extruder quickly at startup. The traditional method of measuring ply thicknesses involves cross-sectioning the laminate and is time-consuming.

Several optical techniques have been used in the past to measure thickness in single-layer films. Interference techniques used to measure thickness or range (as in profilometry) include simple Michelson interferometers (Hartman, 1987), phase unwrapping schemes (Strand & Katzir, 1987), and optical heterodynes (Peterson et al., 1984). An optical range-finding method has been described by Xiao et al. (1987). The system focuses reflected light through a pinhole to achieve a stated range resolution of about 2 μm; the system has a working distance of 15 cm. It is possible to employ this approach to measure the thickness of multiple layers in a laminate because light will be reflected not only from the front surface, but also from the interfaces between the layers. Therefore, the distance between these reflective surfaces will be equal to the thickness of the layers. This concept—measuring apparent distances between interfaces with an optical ranging instrument—is the basis of a simple instrument for measuring the thicknesses of both single films and coextruded layers. This optical technique is nondestructive, fast, inexpensive, and suitable for on-line operation.

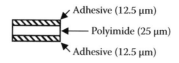

FIGURE 10.13 Cross-sectional view of a three-ply laminate.

10.3.1 Principle of Operation

The measurement technique is based on a confocal optical arrangement that employs ray optics, not interferometry (Figure 10.14). Light from a diode laser passes through a beam splitter and is brought to a focus by a microscope objective lens. If an interface between two media of differing refractive index is present at the focal point, then some of the light will be reflected back into the objective; the beam splitter will send some of the light through a pinhole into a detector. Light reflected from the interface will enter the detector only if the interface is precisely at a predefined distance from the objective lens; otherwise the reflected light will be out of focus at the pinhole. Thus, as a sample is moved into the focal spot of the objective, the response function of the detector will be like that shown in Figure 10.14. A response peak will be obtained for the front surface, the back surface, and each of the interior interfaces within the laminate. In the present case of a three-ply structure, the sensor's response has a total of four peaks; the thicknesses of the three layers are determined by the spatial separation of the four peaks.

The precision of the instrument is determined by the *depth of focus* (more precisely, the *Rayleigh range*) of the optical system. The full width at half maximum (FWHM) of the response curve is equal to the Rayleigh range of the objective lens (see section 10.5 for a full discussion of this point). Theoretically, the FWHM Δ of the response curve is dependent only on the wavelength of the light λ and the numerical aperture (NA) of the objective lens:

$$\Delta = \frac{\lambda}{\pi}\left(\frac{1}{NA}\right)^2 \qquad (10.32)$$

Note that the *NA* used here is the effective numerical aperture of the whole optical system, not the *NA* value shown on the objective lens.

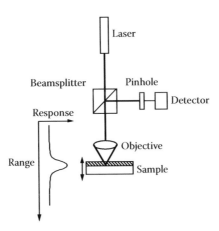

FIGURE 10.14 Optical layout of gauging instrument and its response to an interface where there is a step change in refractive index.

Since the refractive index is higher in the film than in air, for a single ply the apparent distance between the front and back surfaces is shortened. In order to obtain the true thickness t of the first polymer layer, one must measure the optical thickness t' and multiply by a correction factor (derived in section 10.6) that depends on the index of refraction of the first layer:

$$t = \left(\frac{n_1}{n_0}\right)\left[\frac{1-\left(\frac{n_0}{n_1}\right)^2 (NA)^2}{1-(NA)^2}\right]^{(1/2)} \cdot t' \tag{10.33}$$

where n_0 is the refractive index of air, n_1 is the refractive index of the top layer, and NA is the numerical aperture of the objective lens.

A similar situation occurs when one is measuring several layers in a multilayer laminate. The apparent thickness of each layer is distorted by refraction. It is shown in section 10.6 that the true thickness t_m of layer m is obtained by multiplying its apparent thickness t'_m by a correction factor dependent on the refractive index of that layer:

$$t_m = \left(\frac{n_m}{n_0}\right)\left[\frac{1-\left(\frac{n_0}{n_m}\right)^2 (NA)^2}{1-(NA)^2}\right]^{(1/2)} \cdot t'_m \tag{10.34}$$

Thus, by measuring the apparent thickness and applying the correction formula (10.34), one can measure the thickness of individual plies within the coextruded laminate shown in Figure 10.13.

10.3.2   The Optical Sensor and Test Block

An optical sensor was built according to the arrangement shown schematically in Figure 10.14. The light source is an infrared laser diode whose output is collected and collimated with a small lens. This light passes through a beam splitter cube and into the aperture of a 50X, 0.80 NA objective lens. A pinhole (25 micrometers in diameter) and photo detector are mounted on the side of the beam splitter. The entire optical system, complete with a preamplifier for the photo detector, fits into a small aluminum box with dimensions of approximately 2.5 × 4 × 8 cm. The sensor is powered by an external 15-volt supply and provides an output signal proportional to the intensity of light reaching the detector. This output signal is nearly zero unless an interface with a step change in refractive index happens to be at the focal point of the objective lens.

It should be evident that with a collimated light source it is possible to detect the position of the interface by moving the objective lens instead of the sample. Commercial compact disc (CD) players use a similar approach to track the surface of the disc, and the objective lens from a CD player could be used in the optical arrangement shown above.

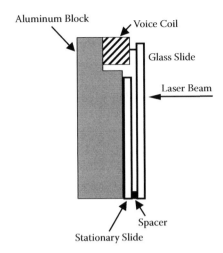

FIGURE 10.15 The testing device used to move samples in and out of the focal place of the instrument shown in Figure 10.14.

In order to evaluate this optical sensor, the testing device shown in Figure 10.15 was used to generate precise small-amplitude motion.[1] The testing device consists of a voice coil removed from a loudspeaker and attached to the end of a glass microscope slide that acts as a cantilever. When a voltage is applied across the voice coil, the glass slide is displaced in proportion to the voltage level. A second glass slide under the cantilever provides a second reflecting surface. Together the stationary slide and the cantilever act as an interferometer, so that the amount of coherent light reflected by the calibration device depends upon the separation between the two slides. The device is calibrated by placing it in the optical system shown in Figure 10.16 and applying a linear ramp voltage to the voice coil. The applied voltage changes the length of the optical cavity formed by the two glass slides; destructive interference occurs when the cavity length is an odd number of quarter wavelengths ($\lambda/4$).

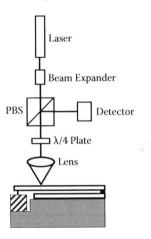

FIGURE 10.16 Optical layout of the interferometer used to calibrate the testing device shown in Figure 10.15.

Figure 10.7 shows the detector output in Figure 10.16 as a function of the ramp voltage. Each cycle in the signal shown in Figure 10.17 represents an interference fringe; the fringes are the result of interference. Using this arrangement one can calibrate the

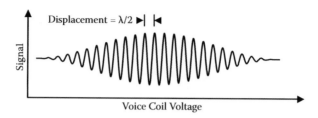

FIGURE 10.17 The reflected light intensity obtained as a function of voice coil voltage in the calibration of the testing device.

voltage applied to the voice coil in terms of the physical displacement imparted to the cantilever. By mounting samples on the cantilever at the same position measured during calibration, it is possible to move the film samples in and out of the focal plane with high precision.

The range resolution of the optical sensor shown in Figure 10.14 was measured by applying a linear ramp signal to the voice coil of the calibrated testing device. The response curve is shown in Figure 10.18, where the full width at half maximum (FWHM) is observed to be about 2.75 μm. This measurement was repeated using microscope objectives of various NA values; the results are compared with theory (equation 10.32) in Figure 10.19. It was observed that the numerical aperture of the optical sensor NA is approximately proportional to the numerical aperture NA' of the objective lens:

$$NA \approx 0.31 NA'$$

(10.35)

For the measurements shown below, the effective numerical aperture of the optical sensor was about 0.25.

10.3.3   Single Film Thickness Measurements

In order to demonstrate the use of this optical technique in measuring film thickness, a series of measurements were made on single Kapton® films ranging from 30 to 500 gauge

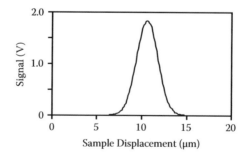

FIGURE 10.18 Detector response to motion of the glass slide in the testing device.

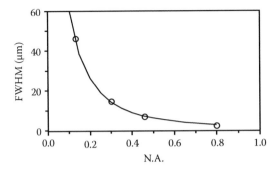

FIGURE 10.19 FWHM as a function of objective NA.

(7.6–127 μm thick). Sample films were fixed to a small washer that was attached to the glass slide on the testing device. The purpose of the washer was to provide a standoff so that the front surface of the slide would not interfere with the back surface of the film. A sketch of the experimental response curve obtained with a sample of 100-gauge polyimide film is shown in Figure 10.20. There are two prominent peaks: one corresponds to the front surface of the film, and the other corresponds to the back surface. The separation of the peaks is the optical film thickness t'; this value is converted into the film thickness t using the correction factor given in equation 10.33.

Seven films of Kapton® type HN polyimide were measured. Gauges studied were 30, 50, 100, 200, 300, 380, and 500 (100 gauge = 0.001 inch = 25.4 micrometers). Each of the films yielded a response curve similar to that shown in Figure 10.20. It was noted that in the case of the thicker samples, the back peak was not as distinct as the front peak. This effect is particularly apparent in the case of the 500-gauge film where the back peak is split into two peaks. The presence of twin peaks is due to birefringence in Kapton®; the effect is most noticeable for thick samples because the separation between the twin peaks is proportional to the total sample thickness. The results of those measurements are shown as circles in Figure 10.21, where optical thickness t' is plotted against nominal film thickness. The linear regression (depicted as a line in the figure) yields

$$t' = (0.5268)t \tag{10.36}$$

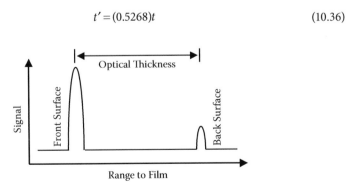

FIGURE 10.20 Detector response versus displacement for a 100-gauge sample of Kapton®.

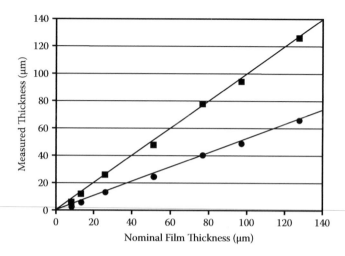

FIGURE 10.21 Apparent thickness (experimental data) versus nominal thickness of polyimide film samples. Circles denote raw data, and squares denote corrected values.

with an R^2 (goodness of fit) value of 0.999. By applying equation 10.33 to this data, the film thickness can be estimated. These measurements are denoted by squares in Figure 10.21, and the linear regression (shown as a line) has an intercept of zero and a slope of 1.0001; therefore the systematic error in these measurements is a mere 0.01%. The experimental data is listed in Table 10.1; in most cases, the individual measurement error is less than 4%.

10.3.4 Application to Multilayer Films

This optical technique can also work on multilayer films, provided the change in refractive index between the layers is sufficiently large. The theory for converting apparent thicknesses to actual thicknesses in multilayer structures is described in section 10.6.

TABLE 10.1 Experimental Data (from Figure 10.21)

Gauge	True Thickness (μm)	Apparent Thickness (μm)	Corrected Thickness (μm)	% Error
30	7.62	3.70	6.85	−10%
50	12.7	6.82	12.8	+0.8%
100	25.4	14.2	26.9	+5.9%
200	50.8	25.7	48.8	−3.9%
300	76.2	41.5	78.9	+3.5%
400	96.5	50.0	95.1	−1.5%
500	127	67(average)	127(average)	0%

When light impinges on an interface between two materials of different refractive index, some of the light is transmitted (refracted) and some is reflected. The reflection coefficient R for normal incidence is given by

$$R = \left(\frac{n_0 - n_1}{n_0 + n_1} \right)^2 \tag{10.37}$$

where n_0 and n_1 are the refractive indices of the two materials (Jenkins & White, 1976). If the difference Δ between the two indices is small, then equation 10.37 simplifies to

$$R \approx \frac{\Delta^2}{4n^2} \tag{10.38}$$

Therefore, the reflection coefficient is proportional to the square of the fractional change (Δ/n). If the fractional change in the refractive index is less than a few percent, the interface becomes difficult to detect, and the individual plies cannot be measured.

If the interface peaks can be observed, then the equations developed in section 10.6 can be used to convert the apparent layer thicknesses into true layer thicknesses. Overall, the method can be expected to perform quite well as long as the appropriate peaks can be observed. Given the difficulty of nondestructively measuring the individual layer thicknesses in multilayer and coextruded films, the simplicity of this technique is quite attractive.

10.4   Appendix on the Thin-Plate Approximation

This section provides the details of the calculations leading up to equations 10.3 and 10.4. The thin-plate approximation worked out here was derived by Lord Rayleigh over a century ago (Strutt, 1945); the following derivation follows a slightly different approach.

Consider a single-frequency, continuous sound wave traveling from left to right and impinging on a very thin plate of infinite extent (the film). The sound wave can be described as a pressure wave moving through the air. This pressure wave moves with a velocity c and obeys the usual wave equation:

$$\frac{\partial^2 P}{\partial x^2} - \frac{1}{c^2} \cdot \frac{\partial^2 P}{\partial t^2} = 0 \tag{10.39}$$

where t in this case is the time variable, not the thickness. For the purpose of this derivation, we will call the thickness τ. The wave equation has a solution

$$P = P_+ e^{i(kx - \omega t)} + P_- e^{-i(kx + \omega t)} \tag{10.40}$$

where the first term represents a pressure wave traveling to the right, and the second term represents a pressure wave traveling to the left. The coefficients P_+ and P_- give the amplitudes of those respective waves, $\omega = 2\pi f$ is the angular frequency of the waves (in rad/s), and k is the wave number defined by $k = \omega/c = 2\pi/\lambda$.

The boundary conditions of this problem are that the pressures and velocities must be the same on both sides of the boundary. In this case, the boundary is defined by the interface between the polymer and the air; therefore there are actually two boundaries, one on each side of the film. The first condition merely requires that the pressure wave given by equation 10.40 be continuous over the boundaries. The second condition, equating velocities on each side, implies a condition on the derivative of the pressure wave.

Consider the pressure differential across a small volume element of thickness dx and cross-sectional area A. This element has a volume $A(dx)$ and is located near the boundary. The pressure differential across the element is given by

$$dP = \left(\frac{\partial P}{\partial x}\right)dx \tag{10.41}$$

where the differential dP is simply the force per unit area acting on the volume element. If the instantaneous velocity of the volume element is v, then its time derivative can be related to dP via Newton's Second Law:

$$F = ma = m\left(\frac{dv}{dt}\right) = \rho(A \cdot dx)\left(\frac{dv}{dt}\right) \tag{10.42}$$

Here, the mass of the element m is expressed as the density ρ of the medium times the volume of the element, and the force is provided by the pressure differential times the area:

$$F = \left(\frac{\partial P}{\partial x}dx\right)A \tag{10.43}$$

Combining equation 10.42 with equation 10.43, we find that

$$\rho\frac{dv}{dt} = \frac{\partial P}{\partial x} \tag{10.44}$$

Now the volume element moves in a harmonic fashion with an angular frequency ω equal to that of the sound wave, so the time part of the velocity goes as $e^{-i\omega t}$; therefore, the first time derivative of v equals a factor $-i\omega$ times the velocity:

$$\frac{dv}{dt} = -i\omega v \tag{10.45}$$

Substituting equation 10.45 into equation 10.44,

$$\therefore v = -\left(\frac{1}{i\omega\rho}\right)\frac{\partial P}{\partial x} \tag{10.46}$$

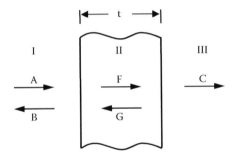

FIGURE 10.22 Cross-sectional view depicting sound transmission through a thin film.

Thus, the requirement that the velocities be the same on both sides of the boundary is really a requirement on the spatial derivative of the pressure wave.

The thin film (or plate) of polymer is shown in cross-section in Figure 10.22, and one can distinguish three regions: region I is the air space on the front surface of the film, region II refers to the interior of the film, and region III is the air space on the back surface of the film. The density ρ_0 and sound speed c_0 of the air are assumed to be the same on both sides of the film, but those properties (ρ_1, c_1) will have different values inside the film. The sound is assumed to be incident from the left (region I), and the boundaries for a film of thickness τ are at $x = 0$ and $x = \tau$. In the absence of reflections, there will be no waves traveling to the left returning from region III, so the pressure wave solution in that region will consist of a single term. From equation 10.40 we can write down the general form of the solution in each of the three regions:

$$\text{Region I:} \quad P_I = Ae^{i(k_0 x - \omega t)} + Be^{-i(k_0 x + \omega t)} \tag{10.47}$$

$$\text{Region II:} \quad P_{II} = Fe^{i(k_1 x - \omega t)} + Ge^{-i(k_1 x + \omega t)} \tag{10.48}$$

$$\text{Region III:} \quad P_{III} = Ce^{i(k_0 x - \omega t)} \tag{10.49}$$

where A, B, C, F, and G are complex amplitudes of the pressure waves (A does not denote area here); $k_0 = \omega/c_0$ is the wave number in air; and $k_1 = \omega/c_1$ is the wave number in the polymer. In particular, A represents the incident wave amplitude, B represents the amplitude of the wave reflected by the film, and C represents the transmitted wave amplitude.

The boundary conditions require these partial solutions to match at $x = 0$ and $x = \tau$. As stated earlier the first boundary condition requires continuity of the pressure waves so that $P_I(x = 0) = P_{II}(x = 0)$ and $P_{II}(x = \tau) = P_{III}(x = \tau)$. Therefore, the first condition yields (after division by the common factor $e^{-i\omega t}$)

$$A + B = F + G \tag{10.50}$$

$$Fe^{ik_1 \tau} + Ge^{-ik_1 \tau} = Ce^{ik_0 \tau} \tag{10.51}$$

The second boundary condition, requiring that $v_I(x = 0) = v_{II}(x = 0)$ and $v_{II}(x = \tau) = v_{III}(x = \tau)$, yields:

$$\frac{1}{i\omega c_0}(ik_0 A - ik_0 B) = \frac{1}{i\omega c_1}(ik_1 F - ik_1 G)$$

$$\therefore \frac{1}{\rho_0 c_0}(A - B) = \frac{1}{\rho_1 c_1}(F - G) \tag{10.52}$$

$$\frac{1}{\rho_1 c_1}(F e^{ik_1\tau} - G e^{-ik_1\tau}) = \frac{1}{\rho_0 c_0} C e^{ik_0\tau} \tag{10.53}$$

where equation 10.46 and the definition of the wave number k have been used. The acoustic impedance z of a material is the product of its density and sound speed, so $z_0 = \rho_0 c_0$ and $z_1 = \rho_1 c_1$. Thus, equations 10.52 and 10.53 become

$$(A - B) = \frac{z_0}{z_1}(F - G) \tag{10.54}$$

and

$$F e^{ik_1\tau} - G e^{-ik_1\tau} = \frac{z_1}{z_0} C e^{ik_0\tau} \tag{10.55}$$

The sum and difference of equations 10.50 and 10.54 yield

$$A = \frac{1}{2}\left(1 + \frac{z_0}{z_1}\right)F + \frac{1}{2}\left(1 - \frac{z_0}{z_1}\right)G \tag{10.56}$$

$$B = \frac{1}{2}\left(1 - \frac{z_0}{z_1}\right)F + \frac{1}{2}\left(1 + \frac{z_0}{z_1}\right)G \tag{10.57}$$

while the sum and difference of equations 10.51 and 10.55 yield

$$F = \frac{1}{2}e^{ik_1\tau}\left(1 + \frac{z_1}{z_0}\right)C e^{ik_0\tau} \tag{10.58}$$

$$G = \frac{1}{2}e^{-ik_1\tau}\left(1 - \frac{z_1}{z_0}\right)C e^{ik_0\tau} \tag{10.59}$$

These four equations (10.56–10.59) are sufficient to solve for the reflection coefficient $R = |B/A|$ and the transmission coefficient $T = |C/A|$.

Recalling that A represents the amplitude of the incident pressure wave, then the reflection coefficient R is the magnitude of the ratio B/A, which upon substitution of equations 10.58 and 10.59 into equations 10.56 and 10.57 can be written explicitly as

$$\frac{A}{B} = \frac{\left(z_1^2 - z_0^2\right)e^{-ik_1\tau} - \left(z_1^2 - z_0^2\right)e^{ik_1\tau}}{\left(z_1 + z_0\right)^2 e^{-ik_1\tau} - \left(z_1 - z_0\right)^2 e^{ik_1\tau}}$$

$$= \frac{-2i\left(z_1^2 - z_0^2\right)\sin(k_1\tau)}{4z_1z_0\cos(k_1\tau) - 2i\left(z_1^2 + z_0^2\right)\sin(k_1\tau)} \tag{10.60}$$

Finding the magnitude of B/A is accomplished by multiplication with its complex conjugate and taking the square root. Thus,

$$R^2 = \left(\frac{B}{A}\right)^*\left(\frac{B}{A}\right) = \frac{4\left(z_1^2 - z_0^2\right)^2\sin^2(k_1\tau)}{16z_1^2 z_0^2 \cos^2(k_1\tau) + 4\left(z_1^2 + z_0^2\right)^2\sin^2(k_1\tau)}$$

$$\therefore R = \frac{\left(z_1^2 - z_0^2\right)\sin(k_1\tau)}{\sqrt{4z_1^2 z_0^2 \cos^2(k_1\tau) + \left(z_1^2 + z_0^2\right)^2\sin^2(k_1\tau)}} \tag{10.61}$$

Until this point the derivation has been exact. Two approximations are now made. The acoustic impedance of a polymer is much greater than that of air, so that $z_1 \gg z_0$ and terms of order z_0^2 can be dropped. We also note that since the polymer sheet is very thin compared to the wavelength of sound in the polymer, $k_1\tau \ll 1$; therefore $\sin(k_1\tau) = k_1\tau$. This second condition, the thin-plate approximation, is justified by the fact that the wavelength in the film is on the order of 10 mm. Thus, equation 10.61 may be approximated as

$$R \approx \frac{z_1^2(k_1\tau)}{\sqrt{4z_1^2 z_0^2 + z_1^4(k_1\tau)^2}} = \left[\frac{4z_0^2}{z_1^2(k_1\tau)^2} + 1\right]^{-1/2} = \left[\frac{(\rho_0 c_0)^2}{(\rho_1 \pi f \tau)^2} + 1\right]^{-1/2} \tag{10.62}$$

where the wave number, acoustic impedance, and angular frequency are replaced by their definitions. With the definition of a constant $\alpha = \pi\rho_1/c_0\rho_0$, the reflection coefficient is written as

$$R = [1 + (\alpha f \tau)^{-2}]^{-1/2} \tag{10.63}$$

This result is given as equation 10.4 in section 10.2.1, where the thickness τ is denoted as t and f is the frequency of the ultrasound (in cycles/second).

The transmission coefficient T, given by the magnitude of the radio $|C/A|$, can be calculated by substituting equations 10.58 and 10.59 into equation 10.56, solving for (C/A), and proceeding as shown above. However, a simpler derivation is based on the fact that since the energy of the ultrasonic wave is conserved, the squares of the reflection and transmission coefficients must sum to unity: $T^2 + R^2 = 1$. By substituting equation 10.63 and solving for T, we find that

$$T = [1 + (\alpha f \tau)^2]^{-1/2} \tag{10.64}$$

This expression for the transmission coefficient T is shown previously as equation 10.3.

10.5 Appendix on Depth of Focus

The purpose of this section is to provide the theoretical justification for the optical technique described in section 10.3.

10.5.1 Laser Beams and Gaussian Optics

Contrary to popular belief, the output beam of a laser is highly collimated but does not remain so (Kogelnik, 1979); the beam starts to spread as soon as it leaves the exit aperture of the laser (Figure 10.23). At some distance z from the laser aperture, the width $W(z)$ of the $1/e^2$ irradiance contour is given by

$$W(z) = w_0 \sqrt{1 + \left(\frac{\lambda z}{\pi w_0^2} \right)^2} \tag{10.65}$$

where λ is the wavelength of the light and w_0 is the waist radius (i.e., the radius of the $1/e^2$ irradiance contour at the plane where the wave fronts are flat). The irradiance distribution $I(r)$ of the lowest-order (TEM00) mode is given by

$$I(r) = \frac{2P}{\pi W^2} \exp \left(-\frac{2r^2}{W^2} \right) \tag{10.66}$$

where $W = W(z)$ as given in equation 10.65, P is the power of the laser, and r is the radius (perpendicular distance from the beam centerline).

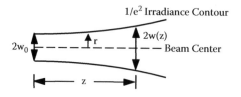

FIGURE 10.23 The divergence of beam at the exit aperture of laser.

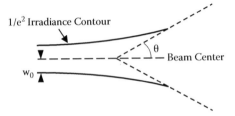

FIGURE 10.24 The equivalent emission cone of a laser.

In the limit of large z, the radius $W(z)$ of the $1/e^2$ contour shown in Figure 10.23 can be approximated by

$$W(z) = \frac{\lambda z}{\pi w_0}$$

(10.67)

The $1/e^2$ contour surface at large z approximates a cone with an apex angle 2θ given by

$$\theta = \frac{W(z)}{z} = \frac{\lambda}{\pi w_0}$$

(10.68)

Therefore, the beam of light emitted by a laser is actually better described as a cone of light, where the apex angle is generally very small. This description is depicted in Figure 10.24.

The so-called *Rayleigh range* Z_r is defined in Figure 10.25 so that $W(+Z_r) = W(-Z_r) = w_0\sqrt{2}$. Setting this condition on the original expression (equation 10.65) for the beam contour, we find that

$$w_0\sqrt{2} = w_0\left[1 + \left(\frac{\lambda Z_r}{\pi w_0^2}\right)^2\right]^{1/2}$$

(10.69)

which requires that

$$1 = \left(\frac{\lambda Z_r}{\pi w_0^2}\right)^2$$

(10.70)

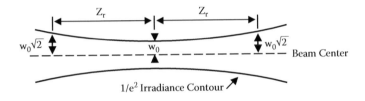

FIGURE 10.25 The definition of the Rayleigh range.

Therefore, solving for Z_r, we find that

$$Z_r = \frac{w_0^2 \pi}{\lambda} \tag{10.71}$$

10.5.2 Beam Concentration

A lens will focus a collimated beam down to a waist at a distance of one focal length f (not to be confused with frequency) from the lens. If we consider the process with time reversed, we see the beam expanding (as it moves away from the focus) to fill the lens, which has an aperture equal to the beam radius $W(z = f)$. If the focal length f of the lens is long enough to satisfy the far-field approximation $z \gg (\pi w_0 / \lambda)$, then the aperture of the lens is given by

$$W = \frac{\lambda f}{\pi w_0} \tag{10.72}$$

Solving for w_0, the waist size at the focus, we find that

$$w_0 = \frac{\lambda f}{\pi W} \tag{10.73}$$

where W is interpreted as the size of the lens aperture. Therefore, the size of the waist at the focus is directly proportional to the focal length of the lens and inversely proportional to the size of the aperture.

10.5.3 Depth of Focus

The term *depth of focus* refers to the sensitivity of the beam size $W(z)$ to variations in the range parameter z. If we choose the depth of focus to be that change (Δ) in z necessary to increase the beam size to $\sqrt{2}$ times the beam waist, then we obtain the following:

$$W(z = \Delta) = w_0 \sqrt{2} \tag{10.74}$$

From equations 10.65 and 10.74 we deduce that

$$w_0 \sqrt{2} = w_0 \left[1 + \left(\frac{\lambda \Delta}{\pi w_0^2} \right)^2 \right]^{1/2} \tag{10.75}$$

which implies

$$1 = \frac{\lambda \Delta}{\pi w_0^2} \tag{10.76}$$

and therefore (Self, 1983),

$$\Delta = \frac{\pi w_0^2}{\lambda} \tag{10.77}$$

Comparing the result of equation 10.77 with that of equation 10.71, we see that the depth of focus Δ defined here is equal to the Rayleigh range Z_r defined earlier. The waist radius w_0 has already been determined in equation 10.73, where W is now the aperture of the lens and f is the focal length. Substituting equation 10.73 into equation 10.77 yields

$$\Delta = \frac{\lambda f^2}{\pi W^2} \tag{10.78}$$

The NA of an objective lens is defined to be

$$NA = \frac{W}{f} \tag{10.79}$$

Therefore, the depth of focus is dependent solely on the wavelength of the light and the NA of the objective lens:

$$\Delta = \frac{\lambda}{\pi} \left(\frac{1}{NA} \right)^2 \tag{10.80}$$

The beam waist at the focus is given by equation 10.73; in terms of the numerical aperture, the waist is given by

$$w_0 = \frac{\lambda}{\pi} \left(\frac{1}{NA} \right) = \Delta \tag{10.81}$$

10.6 Appendix on Thickness Correction

This section discusses the effect that refraction has on the apparent (optical) thickness of the polymer layers in a laminate made using the apparatus shown in Figure 10.14.

In order to measure the thickness of a polymer film, the front and back surfaces are located by changing the separation between the film and the objective lens and noting the two positions at which the maximum signal is detected. The distance between these two positions corresponds to the distance between the front and back surfaces (i.e., the thickness of the film). The distance between the response peaks is in fact only the apparent thickness of the sample. Due to refraction of light in the film, the true thickness is actually greater than the apparent thickness measured by the optical sensor, so a correction factor is needed.

10.6.1 Single-Ply Films

Consider a single-ply film, as shown in Figure 10.26. Assuming a collimated light source, a light ray emanating from the objective lens strikes the front surface at an angle θ_0 that is determined by its effective *NA*:

$$NA = \sin\theta_0 \tag{10.82}$$

The detector response is at a maximum whenever the focal point is at an interface between two regions of differing refractive index. Here the film has a refractive index n_1 and the surrounding air has an index n_0. The apparent thickness t' of the film is simply the physical distance through which one moves the film (or equivalently, the objective) to go from the front surface response peak to the back surface response peak. As the figure illustrates, however, the apparent thickness is less than the true thickness t. Refraction at the front surface causes the rays to bend toward the normal according to Snell's Law:

$$\sin\theta_1 = \left(\frac{n_0}{n_1}\right)\sin\theta_0 \tag{10.83}$$

Therefore, the focus is moved toward the back surface of the film. In the figure, the solid line represents the actual path of the light ray, and the dotted line represents the apparent ray path.

From the figure, we see that both the apparent thickness t' and the true thickness t can be related to the same distance d:

$$t' = \frac{d}{\tan\theta_0} \tag{10.84}$$

$$t = \frac{d}{\tan\theta_1} \tag{10.85}$$

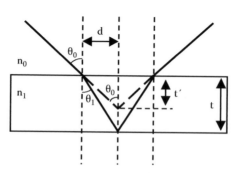

FIGURE 10.26 The geometry used to derive equation 10.90; the focal point of the optical sensor is at the back surface of a single-ply film.

$$\therefore t = \left(\frac{\tan\theta_0}{\tan\theta_1} \right) t' \tag{10.86}$$

The correction factor ($\tan\theta_0 / \tan\theta_1$) in (10.86) relating apparent thickness to true thickness can be derived from equations 10.82 and 10.83:

$$\frac{\tan\theta_0}{\tan\theta_1} = \frac{\sin\theta_0}{\sqrt{1-\sin^2\theta_0}} \cdot \frac{\sqrt{1-\sin^2\theta_1}}{\sin\theta_1} = \left(\frac{n_1}{n_0} \right) \left[\frac{1-\sin^2\theta_1}{1-\sin^2\theta_0} \right]^{1/2} \tag{10.87}$$

$$\frac{\tan\theta_0}{\tan\theta_1} = \left(\frac{n_1}{n_0} \right) \left[\frac{1 - \left(\dfrac{n_0}{n_1} \right)^2 \sin^2\theta_0}{1-\sin^2\theta_0} \right]^{1/2} \tag{10.88}$$

$$\frac{\tan\theta_0}{\tan\theta_1} = \left(\frac{n_1}{n_0} \right) \left[\frac{1 - \left(\dfrac{n_0}{n_1} \right)^2 (NA)^2}{1-(NA)^2} \right]^{1/2} \tag{10.89}$$

Hence,

$$t = \left(\frac{n_1}{n_0} \right) \left[\frac{1 - \left(\dfrac{n_0}{n_1} \right)^2 (NA)^2}{1-(NA)^2} \right]^{1/2} \cdot t' \tag{10.90}$$

The correction factor given by equation 10.89 is a function of the refractive indices as well as the numerical aperture of the lens. For small numerical apertures ($NA < 0.4$), the correction is approximately equal to the ratio (n_1/n_0), so that $t \approx (n_1/n_0)t'$. The depth of focus, and therefore the measurement resolution, is determined by the NA value; therefore it is important to note that equation 10.89 depends on the square of the NA. For the case of a single film with an index $n_1 = 1.766$, the correction factor calculated from equation 10.89 is shown in Figure 10.27.

It should be noted that, in general, the NA is fixed by the optical system, but the index of refraction may change. Therefore, the correction factor will not be the same for every film, and it may change from lot to lot if the film index changes.

10.6.2 Multi-Ply Laminates

In the case of a laminate with distinct layers, each with its own index of refraction, the top layer will have an apparent thickness that can be corrected by equation 10.90. The other layers will also have apparent thicknesses that are altered by refraction.

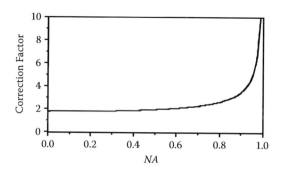

FIGURE 10.27 The correction factor given in equation 10.89, calculated as a function of numerical aperture.

The correction factors for these layers are derived from Snell's Law as in the case of the single film.

The path taken by a light ray from the objective through the top two layers of a multiply laminate is shown in Figure 10.28. The focus is assumed to be at the bottom of the second layer. In the figure, the two distances d_1 and d_2 are related to the apparent thicknesses t'_1 and t'_2 of the first and second layers via the tangent of the incident angle:

$$(d_1 + d_2) = (t'_1 + t'_2) \cdot \tan\theta \tag{10.91}$$

It is also apparent from the figure that

$$d_1 = t_1 \cdot \tan\theta_1 \tag{10.92}$$

and

$$d_2 = t_2 \cdot \tan\theta_2 \tag{10.93}$$

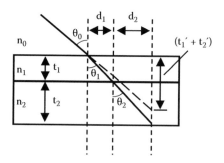

FIGURE 10.28 The geometry used to derive equation 10.97; the focal point of the optical sensor is at the back surface of a two-ply laminate film.

Substituting equations 10.92 and 10.93 into equation 10.91, we find that

$$t_1 \cdot \tan\theta_1 + t_2 \cdot \tan\theta_2 = t_1' \cdot \tan\theta_0 + t_2' \cdot \tan\theta_0 \qquad (10.94)$$

Remembering the result of equation 10.86, it is evident that in equation 10.94 the terms containing t_1 and t_1' are equal to each other. Therefore,

$$t_2 = \left(\frac{\tan\theta_0}{\tan\theta_2} \right) t_2' \qquad (10.95)$$

Following the results obtained in equations 10.87 and 10.88, the ratio of the tangents in equation 10.95 can be expressed in terms of $\sin\theta_0$, recognizing that

$$\frac{\sin\theta_0}{\sin\theta_2} = \frac{\sin\theta_0}{\sin\theta_1} \cdot \frac{\sin\theta_1}{\sin\theta_2} = \frac{n_1}{n_0} \frac{n_2}{n_1} = \frac{n_2}{n_0} \qquad (10.96)$$

Finally, we obtain the correction formula for the true thickness of layer 2:

$$t_2 = \left(\frac{n_2}{n_0} \right) \left[\frac{1 - \left(\dfrac{n_0}{n_2} \right)^2 (NA)^2}{1 - (NA)^2} \right]^{(1/2)} \cdot t_2' \qquad (10.97)$$

Note that the correction is based only on the numerical aperture of the optical sensor and the relative refractive index (n_2/n_0) of the second layer, not the refractive index of the first layer. Similar results are obtained for subsequent layers. In general, the true thickness t_m of layer m is related to the apparent thickness t_m' of that layer via

$$t_m = \left(\frac{n_m}{n_0} \right) \left[\frac{1 - \left(\dfrac{n_0}{n_m} \right)^2 (NA)^2}{1 - (NA)^2} \right]^{(1/2)} \cdot t_m' \qquad (10.98)$$

Given this general result, the true thickness of each layer may be found from the apparent thickness of that layer, the *NA*, and the index of refraction of that layer. The determination of thickness for each layer is therefore independent of the other layers, and measurement errors will not be compounded from one layer to the next.

Suggested Reading

Brekhovskikh, L.M. (1980). *Waves in Layered Media* (translated by RT Beyer). New York: Academic Press.

Krautkramer, J. and Krautkramer, H. (1983). *Ultrasonic Testing of Materials,* Third Edition. New York: Springer-Verlag.

Jenkins, F.A. and White, H.E. (1976). *Fundamentals of Optics,* Fourth Edition. New York: McGraw-Hill.

Strutt, J.W. (Lord Rayleigh), (1945). *The Theory of Sound,* First American Edition. New York: Dover Publications.

11

Plastics and Composite Materials

11.1 Identification of Polymer Type

The economic viability of recycling plastic materials depends on developing inexpensive and fast methods for sorting dirty, crushed plastic bottles and containers. Most sources of recyclable material provide a random mixture of various plastic types, but recycling processes generally require a single polymer to be used. Therefore, the first step of the recycling process is to sort the input waste stream by polymer type.

In many cases this sorting process is still accomplished manually, and it represents a significant portion of the cost associated with the recycling process. In a typical recycling plant, bales of crushed, comingled (that is, comprised of various types), dirty bottles and containers are broken apart and are spread onto conveyor belts for manual separation. These containers are often badly mangled or missing their labels. Accurate separation depends on the ability of unskilled workers to recognize particular types of bottle or container against a background of other types. This approach assumes that the container manufacturer always uses the same plastic resin, an assumption that is sometimes incorrect. Once a bottle has been identified, it is manually tossed onto another conveyor belt carrying bottles of the same type (such as polyethylene terephthalate). The workers generally search for one or (at most) two types of bottle in order to simplify the recognition task.

In the case of polyethylene terephthalate (PET) and polyvinyl chloride (PVC) bottles, the two resins are difficult to distinguish by sight alone. The difficulty is compounded by some manufacturers that indiscriminately use either PET or PVC to make identical bottles for identical products. It is essential to distinguish correctly between these particular polymers because the presence of PVC in the PET remolding process, even at the level of a few parts per million, will ruin the PET resin.

The value of recycled polymers is limited by the cost of the virgin plastics, which in turn is governed by prevailing oil prices. Automating the sorting process can improve profit margins for the existing small-scale recycling operations and is essential for any large-scale effort. A variety of identification methods have been proposed, developed, and commercialized.[1] However, none of the available technologies is very fast, and most of them involve expensive or complicated instruments; clearly, simplified sensors are needed.

This section describes a simple near-infrared sensor that differentiates polymer resins in real time and thus could be used to automate recycling plants (Scott, 1995; Scott & Waterland, 1995).

11.1.1 NIR Absorbance Spectra

The near-infrared (NIR) region of the electromagnetic spectrum extends from about 750 nm to about 2500 nm in wavelength. A variety of chemical bonds (e.g., O–H, C–H) absorb light in this region, and it is well known that an analysis of NIR absorbance can be used to obtain information about the chemical structure of a sample. By measuring the absorption of light at a few well-chosen wavelengths, it is possible to obtain a unique signature for each type of plastic. The signature for an unknown sample can then be compared to those of known plastics in order to determine the type (Scott & Waterland, 1995).

Absorbance at a particular wavelength λ is defined (Stewart, 1970, p. 82)

$$A(\lambda) = -\log_{10}\left[\frac{I(\lambda)}{I_0(\lambda)}\right] \tag{11.1}$$

where $I_0(\lambda)$ is the intensity of light incident on the sample and $I(\lambda)$ is the intensity of the light transmitted through it. The ratio $I(\lambda)/I_0(\lambda)$ is defined to be the transmission coefficient of the sample, so absorbance is determined from the transmission data.

Samples of crushed, dirty plastic were obtained from a commercial recycling plant in Philadelphia. Plastic bottles and containers were removed directly from the sorting chutes and subsequently cut into squares of about 8 cm on a side. Several polymer types were collected, including clear PET, green PET, PVC, natural high-density polyethylene (HDPE) and pigmented HDPE. A total of 59 sample squares were prepared. The samples were not cleaned and, in many cases, included labels or creases. NIR absorbance spectra covering wavelengths from 1100 nm to 2500 nm were measured with an NIR Systems Model 6500 spectrophotometer; examples of these spectra are shown in Figure 11.1 through Figure 11.3.[2]

The spectra collected for various samples of each polymer type proved to be remarkably consistent in spite of the fact that the samples were dirty. This observation suggests

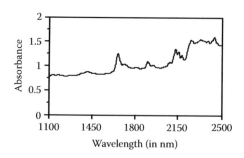

FIGURE 11.1 Absorbance spectrum of green PET. From Scott (1995). Used with permission from the Institute of Physics.

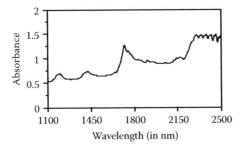

FIGURE 11.2 Absorbance spectrum of PVC. From Scott, 1995. Used with permission from the Institute of Physics.

that typical levels of contamination do not obscure characteristic spectral features. The only variation noted between spectra of a given polymer was in the total absorbance values. Absolute absorbance (which is measured in units of optical density) is directly proportional to the thickness of the sample, so it is useful to consider the ratio of absorbance values at two different wavelengths. To first order, this ratio is independent of the wall thickness of the plastic bottles.

In the region between 1000 nm and 2000 nm, the location and relative height of absorbance peaks were found to be unique for each polymer type. Referring to Figure 11.1, a dominant absorbance peak in PET is at 1660 nm. In PVC (Figure 11.2), this peak has shifted to 1718 nm. This feature has been observed in a number of polymers and is attributed to the first overtone of C-H stretching (Crandall & Jagtap, 1977). PVC also exhibits two small, broad peaks located at 1196 nm and 1422 nm. Figure 11.3 shows a typical spectrum of HDPE, in which there are prominent peaks at 1214 nm and 1732 nm. From these observations, it follows that we can focus on absorbance data near wavelengths of 1214, 1660, and 1718 nm in order to distinguish these three polymer types. It is serendipitous that these wavelengths lie in relatively flat portions of the absorption spectrum of water, because in most cases the plastic will carry some moisture with it (Palmer & Williams, 1974).

Absorbance data can be provided by a simple fixed-filter spectrometer that consists of a light source and several NIR detectors, each with its own optical filter. Such devices are faster, more rugged, and less expensive than the instrument that produced the continuous spectra shown in Figure 11.1 through Figure 11.3. The prototype instrument described in

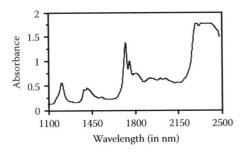

FIGURE 11.3 Absorbance spectrum of pigmented HDPE. From Scott & Waterland, 1995.

section 11.1.3 uses interference filters to select the wavelengths of interest (i.e., 1214, 1660, and 1716 nm). The output signals from the three detectors represent the transmitted light intensity at the three wavelengths. In order to correct for variations in sample thickness, the data was normalized by dividing the readings at 1660 nm and 1716 nm by the reading at 1214 nm. These two ratios were then used as inputs to a neural network, which is described in the following section.

11.1.2 Neural Networks

In 1943 McCulloch and Pitts showed that Boolean functions could be computed using simple "neural" processing elements loosely modeled after biological neurons (McCulloch & Pitts, 1943). This remarkable result marked the birth of a new area of computation that has come to be known as *neural networks*.[3] This discipline has developed into a mature field with important contributions from physics, computer science and neurobiology. New developments in the theory of neural computation have led to impressive real-world applications in process control, process modeling and forecasting.

Neural networks are composed of many simple processors ("neurons") that pass information to one another through weighted connections. Neural network computers are not programmed; rather, they learn about their environment by repeated exposure to examples of desired behavior. Typically, each neuron applies a nonlinear transformation to a weighted sum of its inputs and adapts to its environment according to a learning rule (Rumelhart et al., 1986). The three-layer network shown in Figure 11.4 was used to identify HDPE, PET, and PVC using NIR data as described below.

This neural network has two inputs. The first input I_1 is the ratio of the response at 1660 nm to that at 1214 nm. The second input I_2 is the ratio of the response at 1716 nm to that at 1214 nm. The three outputs of the network can regarded as indicator lights. The goal is to train the system so that the first light turns on if the polymer sample is HDPE, the second light turns on if it is PET, and the third light turns on if it is PVC. The inputs I_1 and I_2 are connected to two "hidden layer" neurons; each hidden neuron has two input connections, giving a total of four connections from the input layer to the hidden layer. These connections have associated weights that are represented by a 2×2 matrix **W1**.

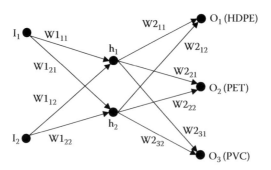

FIGURE 11.4 The three-layer neural network used for identifying polymer type. Adapted from Scott & Waterland, 1995.

Each of the hidden unit neurons computes a weighted sum of its inputs and subtracts a threshold value $\theta 1_i$ to yield an intermediate value g_i:

$$\begin{bmatrix} g_1 \\ g_2 \end{bmatrix} = \begin{bmatrix} W1_{11} & W1_{12} \\ W1_{21} & W1_{22} \end{bmatrix} \begin{bmatrix} I_1 \\ I_2 \end{bmatrix} - \begin{bmatrix} \theta_1 \\ \theta_2 \end{bmatrix} \tag{11.2}$$

If the inputs, thresholds, and intermediate values are represented as vectors, this equation can be written as

$$\mathbf{g} = \mathbf{W1} \cdot \mathbf{I} - \theta\mathbf{1} \tag{11.3}$$

The activity level h_i of each neuron is determined by passing g_i (the corresponding component of $\mathbf{g}$) through a sigmoid nonlinear function defined by

$$S(x) = \frac{1}{1 + e^{-x}} \tag{11.4}$$

so that

$$h_i = S(g_i) \tag{11.5}$$

These values are passed in turn to three output layer neurons through six weighted connections represented as a 3×2 real matrix, $\mathbf{W2}$. The outputs O_1, O_2, and O_3 are the components of the vector O defined by

$$O = S(\mathbf{W2} \cdot \mathbf{h} - \theta\mathbf{2}) \tag{11.6}$$

where the function S is the vector form of equation 11.4. O_1, O_2, and O_3 can be regarded as the probability of a sample being HDPE, PET, or PVC, respectively.

The network in Figure 11.4 was trained by adjusting the weights $\mathbf{W1}$ and $\mathbf{W2}$ so that a given input I_1 and I_2 produced the correct classification. Measured values of I_1 and I_2 for 37 plastic samples were used to train the network. The training data was repetitively presented to the network and the network outputs were compared with the desired output values. The weights $\mathbf{W1}$ and $\mathbf{W2}$ were modified according to a training scheme so that eventually the desired output was obtained with each of the input combinations in the training set (Filkin, 1991).

After this training was completed, the network was tested using 22 additional samples not seen in the training phase. These new samples were all classified perfectly, indicating that the network had learned to distinguish these three types of plastic. A good gauge of a model's performance is the statistical R^2 parameter, which is the ratio of the variance of the model to that of the testing data. If R^2 is 1 the model is perfect; if R^2 is less than 0.5 the model is very poor. For this problem, the trained neural network has $R^2 = 0.9996$.

11.1.3 Hardware Implementation

Figure 11.5 shows a three-filter spectrometer suitable for sorting crushed plastic bottles by polymer type. A broadband source of NIR light illuminates the plastic sample via a fiber-optic bundle. The author has found that a 15-watt tungsten bulb generates sufficient NIR illumination for this purpose (of course, NIR sources are also available commercially). In any case, the color temperature of the source must be stabilized so that the spectral output remains constant. Matching lenses are used to collimate the light before it passes through the plastic sample. Collimating the light allows a separation of several inches between the fiber-optic bundle that illuminates the sample and the bundle that receives the transmitted light.

Light transmitted by the plastic bottle returns to the instrument via a fiber optic bundle that splits into three separate bundles. Each of these smaller bundles is terminated by a filter/detector combination. The 10 nm bandpass interference filters select the wavelengths of interest, so that the corresponding detector measures the light intensity at that wavelength only. Cooled detectors are used to measure the intensity of the NIR light after it passes through the plastic.

The outputs from the three detectors are conditioned by logarithmic amplifiers in order to make the signal proportional to absorbance (see equation 11.1). The amplifier outputs are digitized and recorded by the computer. Absorbance measurements require the intensity $I_0(\lambda)$ of incident light to be recorded for all three wavelengths by measuring

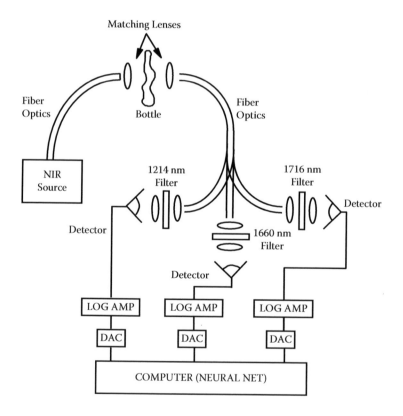

FIGURE 11.5 Block diagram of a fixed-filter NIR instrument used for identifying polymer type. From Scott & Waterland, 1995.

the detector signals without any sample in the light path. Subsequent readings with samples in place are then used to calculate the absorbance. The computer that controls the instrument is programmed to implement the neural network discussed in the previous section. Once the absorbance values have been measured, inputs I_1 and I_2 are calculated from

$$I_1 = \frac{A(1660 \text{ nm})}{A(1214 \text{ nm})} \tag{11.7}$$

$$I_2 = \frac{A(1718 \text{ nm})}{A(1214 \text{ nm})} \tag{11.8}$$

and the neural network output is evaluated. The result determines the disposition of the bottle, which can be routed to the appropriate chute by a mechanical device.

The instrument shown in Figure 11.5 has no moving parts and is very compact. This sensor is inexpensive, compact, and very fast. The advantage of using a neural network is that the system can be trained to recognize new types of plastic. A simplified two-wavelength version has also been demonstrated for separating PVC from PET (Scott, 1995).

11.2   Contamination Detection in Molten Polymer

The manufacture of nylon and other thermoplastic polymers generally involves casting or spinning the molten material following the polymerization cycle. Occasional manufacturing problems arise due to particulate contamination in the polymer, which causes filament breakage during spinning, dyeability variations, and other performance problems. The most common contaminates are gels (lumps of cross-linked polymer), bits of thermally degraded polymer, and extraneous particles entrained in the process during charging of the autoclave (where polymerization takes place).

Polymer quality problems related to particulate contamination are costly in terms of manufacturing resources and customer satisfaction. The ability to detect, reduce or eliminate these particles directly impacts product properties, quality, waste generation, and downtime. Melt phase monitoring of contaminates is a first step to contain losses due to process upsets. When an upset does occur, characterizing the size and number of contaminant particles in the melt facilitates the correct disposition of the product. The root cause of the problem can often be diagnosed by observing the timing of the onset and duration of the upset.

Since the contaminants typically found in molten polymer streams are larger than 10 mm in diameter yet occur at very low (parts per million) levels, process imaging can be used to measure their quantity and size. A variety of optical probes (both commercial and customized) have been developed to detect the presence of gel particles and other contaminants in polymer processes. Some of these probes have been used to study mixing and residence time distributions in extruders. These camera systems must withstand the harsh environment of polymer extruders that operate around 320°C and pressures of about 55 MPa (over 540 atmospheres).

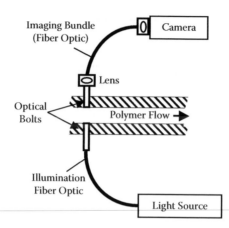

FIGURE 11.6 Block diagram of an imaging system for detecting contamination in molten polymer.

One approach to measuring contamination in molten polymer is to analyze images obtained by transmitting visible light (via fiber optic bundles) through the polymer flow.[4] As shown in Figure 11.6, two opposing "optical bolts" (hollow bolts containing sapphire windows) are installed in the process to provide optical access. Contaminants in the polymer flow block a portion of the light, casting a shadow on the opposite sapphire window. A fiber optic imaging bundle, which preserves the spatial relationship among fibers in the bundle, carries the image to a charge-coupled device (CCD) camera, and the video signal is digitized for image analysis. A region of interest (ROI) is defined within the camera's field of view, and contaminants appearing in that ROI are analyzed using edge detection and other feature extraction techniques. Particle size distributions and total count rates can be automatically generated for the observed particles.

This technology has been used to detect contamination in a production autoclave flake-casting nozzle (Scott, 2005). The autoclave produces nearly 1.4 tonnes (metric tons) of polymer during a two-hour cycle, but casting takes only 18 minutes; the polymer contamination level is monitored during this final casting stage. The count rate data shown in Figure 11.7 demonstrate that the level of contamination can vary considerably

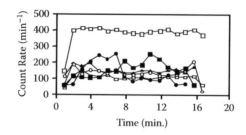

FIGURE 11.7 Contamination count rate as a function of time for several batches of polymer. From Scott (2005).

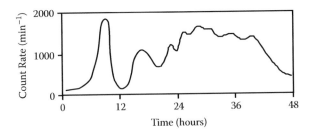

FIGURE 11.8 Contamination count rate as a function of time in a continuous polymerization process. From Scott (2005).

from batch to batch; typically the rate is below 200 counts per minute, but one batch (denoted by open squares in the figure) contained twice as many contaminants. This effect is probably due to pressure fluctuations that jolt degraded polymer residue loose from the autoclave walls and internal piping. By monitoring the level of contamination, manufacturing personnel can identify which production lines need to be cleaned before major problems arise.

Another application is to detect particulate releases in continuous polymerization (CP) operations. CP lines feed spinning cells where the molten polymer is spun into fibers. In such applications, particulate impurities in the polymer decrease the luster of the fiber surface and increase the possibility of thread breakage, which disrupts the manufacturing operation. Contamination increases have been tied to specific events; for example, a sheared pin in a transfer line metering pump caused a major particulate release, as shown in Figure 11.8. The increase in total particulate counts coincided with a 30% loss in fiber luster (measured with a separate in-line sensor), and the upset condition lasted nearly 48 hours. When the particulate contamination decreased to an acceptable level, normal production was resumed. These examples demonstrate how imaging technology can be used to separate off-quality material from normal product automatically.

11.3 Characterization of Reinforced Polymers

11.3.1 Applications of Reinforced Polymers

Reinforced plastic is used for injection-molded parts that require dimensional stability and resistance to chemicals. Filler materials may include chopped glass fibers, glass and mineral combinations, or minerals alone. Polymer resins and blends may also be reinforced with carbon fiber, improving the tensile strength and strength-to-weight ratios.

An important segment of the reinforced thermoplastics market is in the automobile industry, where applications include structural components, body panels, hoods, and engine components. Presently, 50% of the nylon used in light vehicles is reinforced. Reinforced nylon has good impact and abrasion-resistance and increased strength, stiffness, and creep resistance. Reinforced polypropylene is well suited for use in automobile bodies due to its low cost, chemical inertness, stiffness, and heat resistance. Polypropylene can be reinforced with fiberglass or compounded with fillers such as talc, calcium carbonate, and mica. Fiberglass reinforced polyesters are used primarily in engine components such as distributor caps, brush holders, and fans. Likewise, fiberglass reinforced polycarbonate has good electrical properties, impact strength, dimensional stability and rigidity. Auto body components such as front-end panels are formed from glass filled polycarbonate sheets.

There are many examples of reinforced polymers: Zytel® is a glass reinforced nylon featuring high tensile strength, stiffness, and impact resistance. It is used in many automobile applications including front pistons, cooling fans, radiator end tanks, brake fluid reservoirs, brake heat shields, and parking brake cable covers. Zytel® resins are also used in bicycle parts, roller skates, and ice skates. Mineral-filled nylon offers resistance to impact, weather, and road chemicals. It is tougher than fiberglass reinforced nylon, and it is used to make grills, wheel covers, mirror housings, under-the-hood components, and fuel systems. Rynite® resins are made of glass reinforced polyethylene terephthalate which offers dimensional stability, chemical and heat resistance, good electrical properties, and easy fill and melt flow characteristics in the injection molding process.

Reinforced polymers belong to a larger class of materials called *composites*, which are multiphase systems that combine one or more reinforcing fillers in a matrix. Ultimately, the mechanical properties of a composite part depend not only on material properties but also on how well the filler is dispersed throughout the part, how much total filler is present, and (in the case of fiber reinforcement) the length of the fibers (Carling & Williams, 1990). A part with poor dispersion or inadequate loading will have abnormally low mechanical strength. To control the quality of the composite material, it is therefore necessary to control the total amount as well as the dispersion of the filler.

During product development the dispersion and loading of test samples must be quantified in order to assess the success or failure of a new resin or blend. In particular, any processing problems that occur during compounding or injection molding will result in poor performance of the parts. Dispersion and loading measurements can help to differentiate between "off-spec" material and problems introduced during molding. Destructive techniques are generally used to determine loading, which refers to the weight (or volume) of the filler as a fraction of the whole composite. In the case of inorganic fillers, the most common method is to weigh the samples, burn off the polymer in a furnace, wash away the ash, and weigh the inorganic residue; the ratio between this weight and the initial weight is the filler content. This process takes over 30 minutes.

A survey of the literature shows that various authors have tried to characterize reinforced polymers with infrared spectroscopy, ultrasound, microwaves, and ionizing radiation. The underlying concept of these techniques is that a composite material attenuates energy according to a law of mixtures: since each of the component phases of the composite have a characteristic attenuation coefficient, the overall attenuation coefficient of the composite sample is determined solely by the volume or mass fraction of the constituents.

A few commercial instruments are available to determine fiber loading in composites. One system is based on ultrasonic time-of-flight measurements.[5] The speed of sound in these materials is determined by the speed of sound in the constituent phases and the volume fraction of those phases, but a calibration curve must be constructed for each composition. Another system is based on radiation attenuation measurements.[6] Neither of these two systems can be used to determine fiber dispersion, and both are relatively slow. A novel technique described below uses real-time radiography to get results within seconds (Scott, 1994). Image analysis techniques are employed to determine fiber dispersion and loading automatically.

11.3.2 Measurement Principles

When radiation passes through a material, its initial intensity is reduced according to Beer's Law:

$$I = I_0 e^{-\mu t}$$

(11.9)

Here, I_0 is the intensity of the incident radiation beam, I is the beam intensity after traversing the sample, t is the thickness of the sample [cm], and μ is the linear attenuation coefficient of the sample [cm^{-1}]. It should be noted that the value of μ depends on the energy of the X-rays and on the chemical composition and physical state of the sample. The total attenuation of the beam produced by an X-ray tube would require convolving $\mu(E)$ over the energy spectrum of the tube. Nevertheless samples with the same density and chemical composition will have a reproducible, effective value of μ provided the X-ray source is operating under standard conditions of plate voltage and beam current. As long as these operating conditions are maintained, the energy dependence can be justifiably ignored. Beam hardening (the disproportionate removal of lower-energy X-rays as the beam passes through an object) has not been a problem for these polymers.

By measuring I and I_0, we can determine the linear attenuation coefficient of a sample by solving equation 11.9 for μ:

$$\mu = -\frac{1}{t}\ln\left(\frac{I}{I_0}\right)$$

(11.10)

Equation 11.10 can be used to measure the attenuation coefficients for the reinforcing phase and the polymer matrix, denoted as μ_r and μ_m. It is easily shown (see section 11.5) that in a two-phase system such as glass-reinforced nylon, the volume fraction (loading) α of the glass phase can then be determined from μ_s, the linear attenuation coefficient of the composite sample (Scott, 1997):

$$\alpha = \frac{(\mu_s - \mu_m)}{(\mu_r - \mu_m)}$$

(11.11)

If desired, the volume loading α can be converted into the mass loading β:

$$\beta = \frac{\alpha}{[\alpha(1-R)+R]}$$

(11.12)

where R is the ratio of the density of polymer to the density of glass.

Radiography is a familiar technique that is used in medicine, dentistry, and security screening at airports. It produces an image of the internal structure of opaque objects by passing X-rays through the object and recording the spatial variation in the radiation intensity. Photographic film was traditionally used to make the latent X-ray image visible, but radiographic images can also be produced electronically. Regardless of how the radiographic image is generated, each pixel is equivalent to a tiny detector that measures the intensity at that point. Thus, a radiographic system can be used to measure $I(x, y)$ at each point with coordinates (x, y) across the part. By using equation 11.11 the loading $\alpha(x, y)$ can be determined as a function of position; in effect, each gray level value corresponds to a specific level of filler loading. It should be emphasized that even if the resolution of the radiographic system is too low to image individual fiber, the fibers collectively attenuate radiation more than the polymer resin, creating a measurable effect.

As an example, consider the hypothetical low-resolution image shown as a 4×4 matrix in Figure 11.9a. There are a total of 16 pixels in this image, and it is assumed that 100% transmission of the incident radiation (I_0) corresponds to a gray level of 100. Applying equation 11.10 to the data in Figure 11.9a yields another 4×4 matrix of values for μ_s, the attenuation coefficient of the sample; this matrix is seen in Figure 11.9b. Equation 11.11 relates the matrix of μ_s values to a matrix of values for α, the fiber volume fraction of the part (shown in Figure 11.9c). The fractional numbers in Figure 11.9c represent the fiber fraction; if the polymer content is required, then it is necessary only to recognize that for a fully consolidated sample, fiber content plus polymer content equals 100%, so the 4×4 matrix of Figure 11.9c can be subtracted from 1 to obtain a final matrix (Figure 11.9d) showing polymer fraction. The loading of the part as a whole is obtained by averaging the local values of fiber fraction.

In the foregoing discussion, we have assumed that the thickness of the part is known. As long as the part thickness can be measured or estimated, the attenuation coefficient and hence the loading can be determined.

69	69	69	66
69	72	69	66
69	69	69	66
66	66	66	63

(a) Original Image [gray level]

0.371	0.371	0.371	0.415
0.371	0.327	0.371	0.415
0.371	0.371	0.371	0.415
0.415	0.415	0.415	0.459

(b) Local Attenuation [cm^{-1}]

0.10	0.10	0.10	0.15
0.10	0.05	0.10	0.15
0.10	0.10	0.10	0.15
0.15	0.15	0.15	0.20

(c) Filler Volume Fraction [–]

0.90	0.90	0.90	0.85
0.90	0.95	0.90	0.85
0.90	0.90	0.90	0.85
0.85	0.85	0.85	0.80

(d) Polymer Volume Fraction [–]

FIGURE 11.9 Transformation of a hypothetical 4×4 radiographic image (a) into the corresponding maps of (b) linear attenuation coefficient, (c) filler volume fraction, and (d) matrix volume fraction. From Scott (1994).

11.3.3 Hardware Implementation

Figure 11.10 shows the block diagram of a real-time radiography system used to measure dispersion and loading in composite materials. A microfocus X-ray source is mounted above the x-y table[7] which holds the sample tray. This tray is a frame that is uniformly transparent to X-rays at the operating energies, and it may be configured to hold either test coupons or injection-molded parts. An X-ray image intensifier mounted underneath the x-y table converts the radiation passing through the sample into a visible image.[8] A CCD video camera records the image, and the video signal is digitized by a video frame buffer that is connected to a computer. The system is enclosed in a shielded chamber with a safety interlock that shuts off the radiation when the door is opened.

In operation, test coupons or parts are placed on the sample tray, and the X-ray source is energized to 42 kV at 87 mA of current in the tube. These standard operating conditions have been found to be optimal for the glass-reinforced polymer samples at about 3 mm thick, but other settings may be required for other materials. The computer measures each sample sequentially by moving the x-y table to each of the fixed locations on the sample tray.

It is preferable to average many images together in order to reduce the effects of noise in the image intensifier and camera. One should also correct the image for gain and offset variations within the CCD elements of the camera (Scott, 1989). As discussed in section 6.4.2, this correction is made by first recording background and white-level images and then dividing the difference of the raw image and the background image by the difference of white-level and background images. The background image can be measured by putting a lead shutter in front of the beam, whereas the white level image is simply the image obtained with no sample in the beam. The corrected image is analyzed

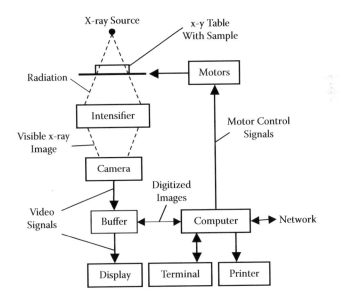

FIGURE 11.10 Schematic diagram of the radioscopic system. Adapted from Scott (1994).

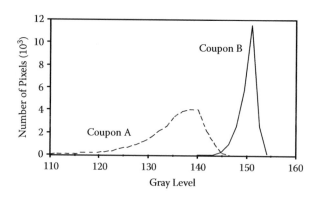

FIGURE 11.11 Histograms of radiographic images. From Scott (1994).

to determine the loading and dispersion. Finally, the computer analyzes the normalized image and generates a report summarizing the measurement data.

11.3.4 Experimental Results

Using the instrument shown in Figure 11.10, radiographs were made of two test coupons of nylon reinforced with glass fiber. One coupon (A) was specifically prepared to have agglomerated fibers; the other one (B) was made to be as uniform as possible. Figure 11.11 shows the histograms of the two images. As expected, the distribution of gray level values is widest for sample A due to clumps of undispersed glass fibers, which appear in the radiograph as small dark blotches. The standard deviation of the gray levels in the image is a measure of the width of the gray level distribution, so it can be used to quantify the dispersion quality. Figure 11.11 also shows the effect of fiber loading on the gray-level distribution: the centroids of the two distributions are not the same because one sample bar (A) contains more glass than the other. Therefore the mean gray level in a region of interest can be related (via equations 11.10 and 11.11) to the loading of the composite in that region.

Figure 11.12 compares loading measurements made with this instrument to the results obtained by the conventional ashing method (Scott, 1997). The samples were composed of both commercial and experimental grades of glass-reinforced Zytel®. Considering that the new technique is based on first principles with no calibration curves, the agreement between predicted and observed values is remarkable. On an absolute scale, there appears to be a systematic error of 2%; however, the relative error is only 0.5%. With greater care, it should be possible to reduce the systematic error. It should also be noted that the ashing method could potentially underestimate the loading if any glass were washed away during the rinsing step. Thus the present measurements may be even closer to the actual values than indicated in the graph.

Figure 11.13 shows the pixel standard deviations for 13 sample lots, where each lot had 7 to 14 test coupons and represented different process conditions for the same experimental composite of glass fiber and polymer. Certain process conditions yielded

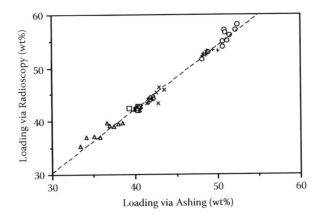

FIGURE 11.12 Loading measured by radiographic system versus loading measured by the destructive ashing method. From Scott (1994).

relatively good fiber dispersion (indicated by standard deviations that were both reproducible and small), whereas many process conditions did not. For comparison, it should be mentioned here that commercial grades of glass-reinforced nylon typically have a standard deviation of about 0.8.

The instrument shown in Figure 11.10 was used to examine over 5,000 samples made from 14 types of polymer. These measurements were used to guide improvements to the extruder design, choice of fiberglass, and compounding conditions. Melt viscosity, a factor suspected to be important to fiber dispersion, was in fact shown to correlate with the dispersion data measured with this instrument. Measurements of standard deviation

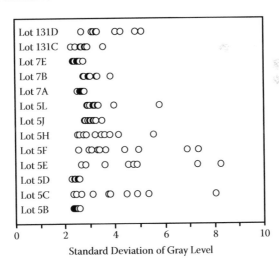

FIGURE 11.13 Process conditions affect the uniformity of the reinforced polymer, as shown here by the standard deviation of gray level values in the radiographs. From Scott (1994).

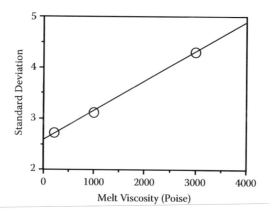

FIGURE 11.14 Observed standard deviation versus melt viscosity. From Scott (1997). ©Elsevier; used with permission.

versus melt viscosity for three resins are shown in Figure 11.14. The improved dispersion at lower melt viscosity is assumed to be the result of better wetting of glass fibers by the resin. Also, polymer resins that contained plasticizer generally exhibited better glass dispersion due to the reduced viscosity. Other processing details such as glass fiber length, screw design, and blending method were also observed to affect the dispersion.

11.4   Measurement of Part Dimensions Using Radioscopy

Conventional radiography images the features of a three-dimensional object by rendering a two-dimensional shadow that is captured on film. A single ray of radiation is attenuated along its flight path from the X-ray source to the film, and its intensity is recorded at a single point on the film. The collection of these individual points comprises the radiograph. A radiograph shows internal features in an object, but it loses the depth information (the distance of the feature from the plane of the film).

If this information is required, there are two classic techniques for measuring depth. The first technique uses two orthogonal side views to determine the distance of the feature from all three sides of the object. The second technique requires the operator to move the X-ray tube a known distance between exposures (Quinn & Sigl, 1980, p. 150ff). Using the parallax (the shift in apparent position of the feature) associated with the movement of the X-ray source, one can determine the flaw depth from the source-to-object-to-film geometry. This well-known technique is depicted in Figure 11.15. Both of these traditional techniques are time-consuming because they require two exposures and the subsequent development of radiographic film.

The technique described below uses the real-time radiography (radioscopy) system shown in Figure 11.10 to measure the parallax associated with motion of the part. This approach allows depth measurements to be made in a matter of seconds. A closely related technique can be used to measure wall thickness in plaques and fabricated parts.

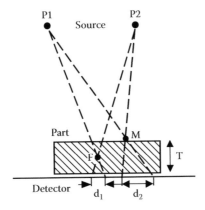

FIGURE 11.15 Film-based parallax measurements. Here $P1$ and $P2$ represent the two positions of the X-ray source; M is a lead marker placed on the surface of the part, T is the part thickness, and F is the flaw. The distances d_1 and d_2 show the parallax of the flaw and marker, respectively. The shift of the flaw image is proportional to the flaw's distance from the film plane; therefore, the flaw height above the film is given by $T(d_1)/(d_2)$.

11.4.1 Concept

The geometry of parallax measurements is shown in Figure 11.16. For the sake of clarity, we will assume that a flaw has been identified in the radiograph of a part and that the operator wants to measure the depth of this flaw. Referring to the figure, a microfocus X-ray source is located at the apex of a triangle defined by the motion of the flaw in the object plane. The position of the flaw's shadow in the detector plane clearly depends upon its position as well as the relative positions of the object and detector planes. The extent

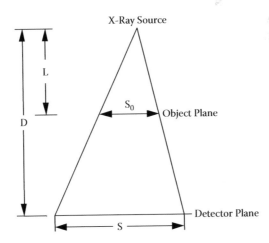

FIGURE 11.16 Parallax geometry.

of the shadow's motion defines a second triangle. Since these triangles are similar, it is clear that

$$L = D\left(\frac{S_0}{S}\right) \tag{11.13}$$

where D is the source-to-detector distance, S_0 is the extent of object motion, S is the extent of the image motion, and L is the distance of the flaw plane from the source. By choosing a suitable reference point on the part (preferably the top surface), L is easily converted into the depth of the flaw. In this system, the part rests upside down on an x-y table that is relatively transparent to X-rays. Since the table surface is at a known, fixed distance from the source, the value of L calculated from equation 11.13 can be subtracted from this constant to yield the depth of the flaw (relative to the surface that rests on the x-y table).

11.4.2 Implementation

This technique has been implemented on the radioscopic imaging system shown in Figure 11.10 (Scott, 1989). When the operator sees a feature or a flaw in the composite part, he can invoke the parallax measurement routine. The computer initially draws a small box on the left and right sides of the image. The operator uses a joystick to move the table so that the flaw image is in one box, then he hits the terminal spacebar. Next, he moves the table so that the flaw image is in the other box, and he hits the spacebar again. The approximate depth of the flaw is displayed and the operator is given the option to proceed. As a practical matter, this first approximation is generally found to be with 1 mm of the final value. This step may be omitted if the sample is not very thick.

If greater accuracy is needed, the operator proceeds with the measurement: the computer records the image of the part with the flaw on one side of the screen, then it moves the part a precise distance toward the other side of the screen. The new image is overlaid on the old image, effectively giving a double exposure. Note that the computer has already determined S_0 by the automatic motion of the table. The operator uses the mouse to define a small ROI containing the flaw image; he then drags it across the display to match it up with the same flaw image in the other location. Clicking the mouse button starts the correlation routine.

The computer calculates a correlation coefficient for the two images in that relative position. A large coefficient signifies that the two images are aligned. The relative position between the two images is automatically varied until the maximum correlation coefficient is obtained. The computer then calculates S, the total distance that the flaw image has moved from its initial position. Finally, equation 11.13 is used to determine the depth of the flaw. It should be noted that generally the flaw has a finite thickness. Therefore the flaw image at the two vantage points will be slightly different due to parallax. By correlating the digitized images, the best overall alignment is achieved.

11.4.3 Depth-of-Flaw Measurements

The entire process described above is menu-driven and can produce accurate results in less than 10 seconds. Figure 11.17 shows actual data taken with this system. A plastic test

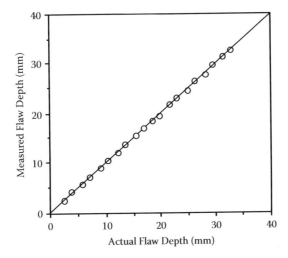

FIGURE 11.17 Depth of flaw data. Open circles are measured values; the line represents a linear fit to the data.

block was prepared by milling 0.8 mm holes at precise locations along the side. These holes, placed at different heights in relation to the x-y table, were imaged with the radioscopic system, and the depth of each hole was determined. The agreement between measurement and nominal location is excellent, as shown in Figure 11.17. The open circles are the measured values, and the line is the linear regression for this data. The offset of this line is negligible, and the slope is 1.0005. Therefore, the systematic error due to calibration is only 0.05%. The total error in the data is typically under 1%, with absolute errors on the order of 0.07 mm or less. The uncertainty of these measurements is due in part to the placement of the holes in the plastic block. The nominal location is known only to within an estimated 0.05 mm.

An error analysis of equation 11.13 yields the equation

$$\left(\frac{\Delta L}{L}\right) = \left[\left(\frac{\Delta D}{D}\right)^2 + \left(\frac{\Delta S_0}{S_0}\right)^2 + \left(\frac{\Delta S}{S}\right)^2\right]^{1/2} \tag{11.14}$$

where $(\Delta L/L)$ is the fractional (relative) error of the measurement. The fractional error $(\Delta S_0/S_0)$ of the object motion is quite small, since the x-y table is controlled by stepper motors with precision lead screws. The estimated absolute error is 0.05 mm; over a travel distance of 5 cm the relative error is only 0.1%. The fractional error $(\Delta D/D)$ in the source/detector separation is also small, because D is large compared to the other distances. This error term is estimated to be about 0.1%. Thus, the greatest contribution to the total error comes from the image resolution of the radioscopic system used to measure S.

The output of the image intensifier in Figure 11.10 is coupled to a CCD video camera. Although the CCD array has a 512×480 pixel format, some of the active area is not useful. A vignetting effect obscures the radioscopic image near the edges of the intensifier's field of view, and the circular field of view is cropped by the rectangular format of the imaging system. Therefore, the useable video images are confined to a square 400×400 pixel area centered on the intensifier output. In order to minimize the error in the measurement of the flaw image, it is necessary to maximize S. Typically, the object motion will be such that the distance S is equivalent to approximately 300 pixels. The error ΔS will be on the order of 1 pixel, so that the relative error is about 0.33%. This uncertainty in S accounts for much of the total error in the measurements presented in Figure 11.17.

11.4.4 Wall Thickness Measurements

In the case of horizontal walls, the procedure described above is easily adapted to measure wall thickness. Figure 11.18 shows the geometry required for this measurement. High contrast markers are affixed to opposite sides of the wall; the exact placement of the markers is not critical. For the case where the wall is horizontal to the detector plane and the direction of motion, it can be shown that the thickness t is

$$t = DS_0 \left(\frac{1}{S_2} - \frac{1}{S_1} \right) \tag{11.15}$$

where D is the source/detector separation, S_0 is the extent of object motion, and S_1 and S_2 are the parallax distances (at the detector plane) of the shadows of the top and bottom markers, respectively.

The technique is slightly more complicated in the case of walls that are pitched at an angle with respect to the direction of motion. A circular marker made of lead foil is placed on the lower surface of the wall, and a second marker is placed on the upper surface. When the circular marker is viewed with the radioscopic system, the tilt of the wall will make the circle appear as an ellipse (Figure 11.19). The slight geometrical distortion of the ellipse due to parallax may be minimized by moving the marker to the

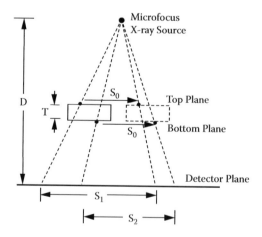

FIGURE 11.18 Wall thickness measurements (horizontal case).

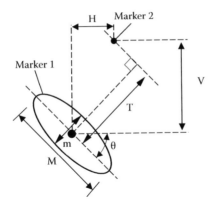

FIGURE 11.19 Wall thickness measurements (tilted case). The circular foil marker is denoted as marker 1. The wall tilt causes marker 1 to appear as an ellipse, with major and minor axes as shown. The vertical distance V is measured between the center of marker 1 and marker 2. The horizontal distance H is measured along the minor axis of the ellipse (which coincides with the direction of the slope).

center of the X-ray beam (indicated by crosshairs on the display screen). The major and minor axes of the ellipse can be measured with an image processing routine; the tilt θ of the wall with respect to the horizontal is given by

$$\theta = \arccos\left(\frac{m}{M}\right) \qquad (11.16)$$

where m is the length of the minor axis and M is the length of the major axis. Next, equation 11.15 is used to determine the vertical distance V between the two markers. Using the crosshairs, the horizontal distance H (Figure 11.19) between the markers is determined by noting the x-y table positions needed to align both markers with the center of the X-ray beam. Finally, the wall thickness t can be calculated from

$$t = V\cos\theta + H\sin\theta \qquad (11.17)$$

Using this technique, it is possible to measure the thickness of walls and other internal dimensions during routine radioscopic inspection of a composite part.

11.5 Appendix on the Calculation of Loading

Equation 11.11 can easily be derived by considering the diminishing intensity of radiation passing through a composite sample. The left side of Figure 11.20 shows a cross-sectional view of a sample of nylon reinforced with fiberglass. For clarity, the glass fibers are shown normal to the cross-sectional plane, but it should be understood that a real sample will have fibers in all possible orientations. As radiation passes through the composite, it is attenuated by both polymer and glass, so that the total transmitted

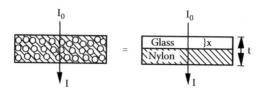

FIGURE 11.20 Cross-sectional view of glass reinforced nylon composite.

radiation is given by Beer's Law (equation 11.9). The order in which the radiation encounters the two phases clearly does not affect the final result, so one can imagine pushing all the glass to the top of the sample (as shown in the right side of Figure 11.20) without changing the radiation transmission measurement.

With the two phases separated (in principle), the incident radiation is attenuated first by an equivalent thickness x of glass, then by a thickness $(t - x)$ of nylon. The glass thickness x is found by considering the glass to be uniformly spread over the same area (A) as the nylon. Therefore the total volume of the glass is (Ax) and the total volume of the nylon is $A(t - x)$. If the volume loading of the glass is denoted as α, then we have

$$\alpha = \frac{Ax}{[Ax + A(t-x)]} = \frac{x}{t} \tag{11.18}$$

so that

$$x = \alpha t \tag{11.19}$$

The equivalent thickness of glass is given by the product of the volume loading α and the thickness t of the sample. It follows that the equivalent thickness of the nylon is $(1 - \alpha)t$.

If the linear attenuation coefficients for glass and nylon (respectively denoted as μ_r and μ_m) are known, then the attenuation due to each phase may be calculated separately from equation 11.9:

$$I_r = I_0 \exp(-\mu_r \alpha t) \tag{11.20}$$

$$I_m = I_r \exp[-\mu_m (1-\alpha)t] \tag{11.21}$$

where I_r and I_m are the intensities transmitted by the equivalent layers of glass and nylon, respectively. Since it is claimed that segregation of the two phases does not change the final transmitted intensity, we can assert that $I = I_m$. It follows from equations 11.9, 11.20, and 11.21 that

$$\mu_s t = \mu_m (1-\alpha)t + \mu_r \alpha t \tag{11.22}$$

where μ_s is the effective linear attenuation coefficient for the composite material. Canceling factors of t and solving for α, we find that

$$\alpha = \frac{(\mu_s - \mu_m)}{(\mu_r - \mu_m)} \qquad (11.23)$$

which is the result given previously as equation 11.11. Therefore the volume loading can be determined by measuring the linear attenuation coefficient of the composite and both phases.

The foregoing analysis has ignored that fact that the linear attenuation coefficient is dependent on the energy of the radiation passing through the sample. For a given atomic element, the attenuation tends to decrease with increasing photon energy; there are however characteristic energies (*K-edges*) at which the attenuation increases significantly. For a polychromatic source such as an X-ray tube, the radiation will be filtered by the sample, effectively convolving the tube's output spectrum with the energy-dependent attenuation $\mu(E)$. In addition, the function $\mu(E)$ for a polyatomic sample is determined by an average over each element in the sample. Nevertheless, samples with the same density and chemical composition will have a reproducible, effective value of μ provided the X-ray source is operating under standard conditions of plate voltage and beam current. As long as these operating conditions are maintained, the energy dependence can be justifiably ignored.

Notes

Chapter 2

1. If the data are not normally distributed (i.e., they do not follow the probability distribution of equation 1.5), then Chebyshev's inequality states that at least 94% of the data fall within 4σ of the mean value. This result means that a deviation must be twice as large (compared to the case of normally distributed data) in order to be considered statistically significant.
2. See the discussion of current, resistance, and voltage in chapter 5.
3. A *boxcar average* (also called a *moving average*) is defined as the mean of the last N readings.

Chapter 3

1. This unit is named after Heinrich Hertz, who demonstrated the existence of radio waves.

Chapter 4

1. The accepted value is close to 2.998×10^8 m/s, but rounding it up to 3×10^8 m/s simplifies the math and introduces only 0.1% error.

Chapter 5

1. A discussion of valence band electrons and conduction band electrons would take us too far off topic. Interested readers should consult a textbook on solid state physics, such as Kittel (2005).
2. Interested readers should consult Halliday et al. (2005), chaps. 28–29.
3. See the discussion in Horowitz and Hill (1980) pp. 25–29.
4. The p-type region is sometimes described as having "holes" that can be filled by electrons in a process called *recombination*. Until the holes are filled, they act as a positive charge carrier.
5. In a cathode-ray tube, these colors are provided by different phosphors that glow red, blue or green when struck by the electron beam; color is determined by modulating the intensity of the electron beam as it sweeps across the screen. Plasma and field-emission displays use different methods of addressing each pixel, but they, too, excite three types of phosphor to create color.
6. See, for example, the discussion in Horowitz and Hill (1980), chapter 3.

7. This digitizer is an AD1671 IC from Analog Devices of Norwood, Massachusetts.
8. For example, the AD12401 is a 12-bit DAC that digitizes at 400 MHz.

Chapter 6

1. The orbital velocity of the electron in a hydrogen atom, for example, is roughly 2.2×10^6 m/s (about 1/137 the speed of light in vacuum).
2. See, for example, Feynman et al. (2006), vol. 3, Lecture 19.
3. The interested reader can find more information on these details in Griffiths (1995) or Winter (1986).
4. It is also possible for an electron to become weakly bound to a neutral atom, thereby creating an *anion*, which is a negative ion. Anions are not considered here.
5. Microfocus tubes with spot sizes on the order of about 10 μm or smaller are available for demanding imaging applications.
6. Large crystals of sodium iodide and certain plastics are often used as scintillator material.
7. L_{max} is generally determined by the number of bits used to represent the gray levels in the image. Many older imaging systems use 8 bits, where L_{max} is 255; newer systems use 10 bits ($L_{max} = 1023$) or more to store the gray level.

Chapter 7

1. The absence of heat does not imply the absence of energy; at 0°K quantum mechanical systems still have a zero-point energy.
2. See, for example, Baker (2000), chap. 12.

Chapter 8

1. The term *Coulter counter* specifically means a device that uses the electrical counting method; it should not be confused with other PSD instruments sold by Beckman-Coulter.
2. The FBRM is available from Mettler Toledo of Columbus, Ohio.
3. See Mie (1908). A comprehensive treatment of the problem is given in Kerker (1969).
4. Some SLS instrument designs place a lens before, rather than after, the sample cell.
5. See for example Kerker (1969) chap. 3 or Ishimaru (1997) chap. 2. Computer programs for calculating the Mie scattering functions are given in the appendix of Bohren and Huffman (1983).
6. DLS can also determine molecular weight and radius of gyration in polymers and proteins.
7. The equations are too complex to be included here; see Challis et al. (1998).
8. As a reminder, the magnitude of a complex number $(a+ib)$, where i is the imaginary number, is given by $\sqrt{a^2+b^2}$. Strictly speaking, the decibel scale measures the signal

amplitude A_1 with respect to a reference level A_2. If $A_1 > A_2$, the signal gain equals $20\log_{10}(A_1/A_2)$; if $A_1 < A_2$, the gain becomes negative, in which case the attenuation in decibels is defined to be $-20\log_{10}(A_1/A_2)$.

9. Although the sound intensity fluctuates wildly along the axis of the transducer in the near field, at high frequencies the spacing between these fluctuations is so fine that the receiving transducer measures an average value of ultrasonic amplitude. See section 10.2.4.

10. The application of high power ultrasound is called *sonication*. A typical power level is 50 watts at a frequency of 40 kHz. By contrast, the ultrasonic power level of the sensor shown in Figure 8.21 is about 1 milliwatt.

Chapter 9

1. Mettler-Toledo (Columbus, OH), J. M. Canty (Buffalo, NY), and Sympatec GmbH (Clausthal-Zellerfeld, Germany) have introduced similar types of in-process imaging probes.
2. Retsch Technology GmbH (Haan, Germany), Microtrac (Montgomeryville, PA), and J. M. Canty (Buffalo, NY) offer particle characterization instruments based on image analysis.
3. This software was developed by Gregg Sunshine.
4. LabView software is sold by National Instruments (Austin, TX).
5. This ECT system comes from Process Tomography Ltd. (Wilmslow, Cheshire, U.K.).

Chapter 10

1. This device was provided by Jerry Lee.

Chapter 11

1. ASOMA Instruments Inc. sells a sensor (ASOMA 652-D) suitable for detecting PVC or other chlorine-bearing polymers. Automated Industrial Controls Inc. has developed a Fourier-transform infrared spectroscopy system for sorting polymer materials.
2. These NIR spectra were measured by Anne Brearley at DuPont.
3. An excellent introductory text is Hertz et al., 1991.
4. The example described here is the Kayeness FlowVision from Dynisco LLC (Franklin, MA).
5. The system is the URA 2000 from TEST Inc. (San Diego, CA).
6. The system is the Compuglass 200 from Radiation Monitoring Devices (Watertown, MA).
7. The source is a model KM10005S microfocus X-ray from Kevex (Scotts Valley, CA).
8. The system is a Dynascope 22 image intensifier, model S41026, from Machlett Laboratories (Stamford, CT).

References

Chapter 2

Bevington, P. R., & Robinson, D. K. (1992). *Data reduction and error analysis for the physical sciences* (2nd ed.). New York: McGraw-Hill.

Fraden, J. (1996). *Handbook of modern sensors*. New York: Springer-Verlag.

Triola, M. F. (2005). *Elementary statistics* (9th ed.). Boston: Pearson/Addison-Wesley.

Chapter 3

Feynman, R. P., Leighton, R., & Sands, M. (2006). *The Feynman lectures on physics: Definitive edition* (vol. 1). Reading MA: Addison-Wesley.

Halliday, D., Resnick, R., & Walker, J. (2005). *Fundamentals of physics* (7th ed., chaps. 16–17). New York: Wiley.

Krautkramer, J., & Krautkramer, H. (1983). *Ultrasonic testing of materials* (3rd ed.). New York: Springer-Verlag.

Povey, M. J. W. (1997). *Ultrasonic techniques for fluids characterization*. San Diego CA: (Academic Press).

Strutt, J. W. (Lord Rayleigh) (1945). *The theory of sound* (1st American ed.). New York: Dover.

Chapter 4

Einstein, A. (1965). Concerning an heuristic point of view toward the emission and transformation of light (trans. A. B. Arons & M. B. Peppard from Annalen der Physik paper of 1905). *Am. J. Phys., 33,* 367–374.

Einstein, A., & Infeld, L. (1966). *The evolution of physics*. New York: Simon and Schuster.

Feynman, R. P., Leighton, R. B., & Sands, M. (2006). *The Feynman lectures on physics: Definitive edition* (vol. 2). Reading MA: Addison-Wesley.

Goodman, J. W. (2005). *Introduction to Fourier optics* (3rd ed.). Greenwood Village CO: Roberts.

Haber-Schaim, U., Dodge, J. H., Gardner, R., Shore, E. A., & Walter, F. (1991). *PSSC Physics* (7th ed., chaps. 17–19). Dubuque, IA: Kendall/Hunt.

Halliday, D., Resnick, R., & Walker, J. (2005). *Fundamentals of physics* (7th ed.). New York: Wiley.

Land, E. H. (1951). Some aspects of the development of sheet polarizers *J. Opt. Soc. Am., 41,* 957–963.

Land, E. H., & Friedman, J. S. (1933). *Polarizing refracting bodies*. U.S. Patent 1,918,848.

Sambursky, S. (1958). Philoponus' interpretation of Aristotle's theory of light. *Osiris, 13,* 114–126.

Yariv, A. (1985). *Optical electronics* (3rd ed., chaps. 4–6, 11, 15). New York: Holt, Rinehart and Winston.

Chapter 5

Halliday, D., Resnick, R., & Walker, J. (2005). *Fundamentals of physics* (7th ed., chaps. 21–30). New York: Wiley.

Horowitz, P., & Hill, W. (1980). *The art of electronics.* New York: Cambridge University Press.

Institute of Electrical and Electronics Engineers. (1985). *IEEE Standard for binary floating-point arithmetic.* IEEE Standard No. 754-1985. New York: Author.

Jackson, J. D. (1975). *Classical electrodynamics* (2nd ed.). New York: Wiley.

Kittle, C. (2005). *Introduction to solid state physics* (8th ed.). Hoboken, NJ: Wiley.

Purcell, E. M. (1985). *Electricity and magnetism* (2nd ed.). New York: McGraw-Hill.

Yariv, A. (1985). *Optical electronics* (3rd ed., chap. 11). New York: Holt, Rinehart, and Winston.

Chapter 6

Feynman, R. P. (2006). Lecture 19: The hydrogen atom and the periodic table. In Feynman, R. P., Leighton, R. B., and Sands, M., *The Feynman lectures on physics: The definitive edition* (vol. 3). Reading, MA: Addison-Wesley.

Griffiths, D. J. (1995). *Introduction to quantum mechanics.* Englewood Cliffs, NJ: Prentice Hall.

Scott, D. M. (1989). Density measurements from radioscopic images. *Materials Evaluation, 47,* 1113–1119.

Sears, F. W., & Zemansky, M. W. (1955). *University physics* (2nd ed.). Reading, MA: Addison-Wesley.

U.S. Code of Federal Regulations. (2007). Title 29, part 1910. Washington, DC: Government Printing Office.

Winter, R. G. (1986). *Quantum physics,* (2nd ed.). Colorado Springs CO: IPI Press.

Chapter 7

Asher, A. C. (1997). *Ultrasonic sensors for chemical and process plant.* Bristol, England: Institute of Physics Publishing.

Baker, R. C. (2000). *Flow measurement handbook: Industrial designs, operating principles, performance, and applications.* New York: Cambridge University Press.

Bernard, M., & Collet, E. (2000). *Hot-wire mass flowmeter.* U.S. Patent 6,035,726.

Eggins, B. R. (2002). *Chemical sensors and biosensors.* Hoboken, NJ: Wiley.

Feynman, R. P., Leighton, R. B., & Sands, M. (2006). *The Feynman lectures on physics: Definitive edition* (vol. 1). Reading MA: Addison-Wesley.

Fraden, J. (2004). *Handbook of modern sensors; Physics, designs, and applications* (3rd ed.). New York: Springer-Verlag.

Halliday, D., Resnick, R., & Walker, J. (2005). *Fundamentals of physics* (7th ed.). New York: Wiley.

Horowitz, P., & Hill, W. (1980). *The art of electronics.* New York: Cambridge University Press.

Kohler, F. (1966). *Resistance element.* U.S. Patent 3,243,753.

Lipták, B. G. (Ed.). (2003). *Instrument engineers' handbook.* Vol. 1, *Process measurement and analysis* (4th ed.). Boca Raton FL: CRC.

Matsuoka, T., Fujimura, M., & Matsuo, Y. (1976). *PTC thermistor composition and method of making the same.* U.S. Patent 3,962,146.

McMillan G. K., & Considine, D. M. (Eds.). (1999). *Process/industrial instruments and controls handbook* (5th ed.). New York: McGraw-Hill.

Sears, F. W., & Zamansky, M. W. (1955). *University physics.* Reading, MA: Addison-Wesley.

Upp, E. L., & LaNasa, P. J. (2002). *Fluid flow measurement: A practical guide to accurate flow measurement* (2nd ed.). Boston: Gulf Professional.

Yariv, A. (1985). *Optical electronics* (3rd ed.). New York: Holt, Rinehart, and Winston, New York.

Chapter 8

Allegra, J. R. (1970). *Theoretical and experimental investigation of the attenuation of sound in suspensions and emulsions.* PhD diss., Harvard University.

Allegra, J. R., & Hawley, S. A. (1972). Attenuation of sound in suspensions and emulsions: Theory and experiments, *J. Acoust. Soc. Am., 51*, 1545–1564.

Allen, T. (1992). *Centrifuge particle size analyzer.* U.S. Patent 5,095,451.

Allen, T. (1997). *Particle size measurement* (5th ed., vol. 1). London: Chapman & Hall.

Atkinson, C. M., & Kytömaa, H. K. (1993). Acoustic properties of solid-liquid mixtures and the limits of ultrasound diagnostics—I: Experiments. *Trans. ASME J. Fluid Eng., 115*, 665–675.

Bohren, C. F., & Huffman, D. R. (1983). *Absorption and scattering of light by small particles.* New York: Wiley.

Brown, R. (1828). A brief account of microscopical observations made in the months of June, July and August, 1827, on the particles contained in the pollen of plants; and on the general existence of active molecules in organic and inorganic bodies. *Phil. Mag., 4*, 161–173.

Challis, R. E., Povey, M. J. W., Mather, M. L., & Holmes, A. K. (2005). Ultrasound techniques for characterizing colloidal dispersions. *Rep. Prog. Phys., 68*, 1541–1637.

Challis, R. E., Tebbutt, J. S., & Holmes, A. K. (1998). Equivalence between three scattering formulations for ultrasonic wave propagation in particulate mixtures. *J. Phys. D: Appl. Phys., 31*, 3481–3497.

Dukhin, A. S., & Goetz, P. J. (2002). *Ultrasound for characterizing colloids.* Amsterdam: Elsevier.

Einstein, A. (1905). Über die von der molekularkinetischen Theorie der Wärme geforderte Bewegung von in ruhenden Flüssigkeiten suspendierten Teilchen (On the Movement of Small Particles Suspended in Stationary Liquids Required by the Molecular-Kinetic Theory of Heat). *Ann Phys.*, 322, 549–560.

Epstein, P. S., & Carhart, R. R. (1953). The absorption of sound in suspensions and emulsions I: Water fog in air. *J. Acoust. Soc. Am.*, 25, 553–565.

Hadamard, J. (1923). *Lectures on the Cauchy problem in linear partial differential equations.* New York: Yale University Press.

Hipp, A. K., Storti, G., & Morbidelli, M. (2002a). Acoustic characterisation of concentrated suspensions and emulsions 1: Model analysis. *Langmuir*, 18, 391–404.

Hipp, A. K., Storti, G., & Morbidelli, M. (2002b). Acoustic characterisation of concentrated suspensions and emulsions 2: Experimental validation. *Langmuir*, 18, 405–512.

Holmes, A. K., & Challis, R. E. (1993). Ultrasonic scattering in concentrated colloidal suspensions. *Colloids and Surfaces A: Physicochemical and Engineering Aspects*, 77, 65–74.

Holmes, A. K., Challis, R. E., & Wedlock, D. J. (1993). A wide bandwidth study of ultrasound velocity and attenuation in suspensions: Comparison of theory with experimental measurements. *J. Colloid Interface Sci.*, 156, 261–268.

International Organization for Standardization (1998). *Representation of results of particle size analysis—Part 1: Graphical representation.* ISO 9276-1. Geneva.

International Organization for Standardization (1999). *Particle size analysis – Laser diffraction methods – Part 1: General principles.* ISO 13320-1. Geneva.

Ishimaru, A. (1997). *Wave propagation and scattering in random media.* New York: IEEE Press.

Kerker, M. (1969). *The scattering of light and other electromagnetic radiation.* New York: Academic Press.

Lloyd, P., & Berry, M. V. (1967). Wave propagation through an assembly of spheres IV: Relations between different multiple scattering theories. *Proc. Phys. Soc.*, 91, 678–688.

Mie, G. (1908). Beiträge zur Optik trüber Medien, speziell kolloidaler Metallösungen, *Annalen der Physik*, 25, 377–445.

Pendse, H. P., & Sharma, A. (1993). Particle size distribution analysis of industrial colloidal slurries using ultrasonic spectroscopy. *Part. Part. Syst. Charact.*, 10, 229–233.

Povey, M. J. W. (1997). *Ultrasonic techniques for fluids characterization.* San Diego CA: Academic Press.

Preikschat, F. K., & Preikschat, E. (1989). *Apparatus and method for particle size analysis.* U.S. Patent 4,871,251.

Riebel, U. (1992). Ultrasonic spectrometry: On-line particle size analysis at extremely high particle concentrations. In N. Stanley-Wood & R. W. Lines (Eds.), *Particle size analysis.* london: Royal Society of Chemistry.

Riebel, U., & Löffler, F. (1989). The Fundamentals of particle size analysis by means of ultrasonic spectrometry. *Part. Part. Syst. Charact.*, 6, 135–143.

Scott, D. M. (1998). Industrial applications of in-line ultrasonic spectroscopy. In V.A. Hackley and J. Texter (Eds.), *Ultrasonic and dielectric characterization techniques for suspended particles*. Westerville OH: American Ceramic Society.

Scott, D. M. (2003). Characterizing particle characterization. *Part. Part. Syst. Charact.*, *20*, 305–310 (2003).

Scott, D. M. (2006). *Method and apparatus for ultrasonic sizing of particles in suspensions*. U.S. Patent 7,010,979.

Scott, D. M., Boxman, A., & Jochen, C. E. (1995). Ultrasonic measurement of sub-micron particles. *Part. Part. Syst. Charact.*, *12*, 269–273.

Scott, D. M., Boxman, A., & Jochen, C. E. (1998). In-line particle characterization. *Part. Part. Syst. Charact.*, *15*, 47–50.

Sewell, C. J. T. (1910). The extinction of sound in a viscous atmosphere by small obstacles of cylindrical and spherical form. *Phil. Trans. Roy. Soc. London, Series A*, *210*, 239–270.

Strutt, J. W. (Lord Rayleigh). (1871). On the light from the sky, its polarization and colour. *Philos. Mag.*, *41*, 107–120, 274–279.

Strutt, J. W. (Lord Rayleigh). (1945). *The theory of sound* (1st American ed.). New York: Dover.

Twomey, S. (1963). *Introduction to the mathematics of inversion in remote sensing and indirect measurements*. Amsterdam: Elsevier.

Urick, R. J. (1948). The Absorption of sound in suspensions of irregular particles. *J. Acoust. Soc. Am.*, *21*, 283–289.

Waterman, P. C. & Truell, R. (1961). Multiple scattering of waves. *J. Math. Phys.*, *2*, 512–540.

Chapter 9

Barrett, H. H., & Swindell, W. (1977). Analog reconstruction methods for transaxial tomography. *Proc IEEE*, *65*, 89–107.

Castleman, K. (1996). *Digital image processing*. Englewood Cliffs, NJ: Prentice Hall.

Feldkamp, L. A., Davis, L. C., & Kress, J. W. (1984). Practical cone-beam algorithm. *J. Opt. Soc. Am.*, *1*, 612–619.

Herman, G. T. (1980). *Image reconstruction from projections*. New York: Academic Press.

Hoyle, B. S., McCann, H., & Scott, D. M. (2005). Process tomography. In D. M. Scott and H. McCann (Eds.), *Process imaging for automatic control*. Boca Raton, FL: CRC.

Iveson, S. M., Litster, J. D., Hapgood, K., & Ennis, B. J. (2001). Nucleation, growth and breakage phenomena in agitated wet granulation processes: A review. *Powder Technology*, *117*, 3–39.

Kwade, A. (1999). Wet comminution in stirred media mills: Research and its practical application. *Powder Technology*, *105*, 14–20.

Rosenfeld, A., & Kak, A. (1982). *Digital picture processing*. New York: Academic Press.

Scott, D. M. (2005). Applications in the chemical process industry. In D. M. Scott and H. McCann (Eds.), *Process imaging for automatic control*. Boca Raton, FL: CRC.

Scott, D. M., Boxman, A., & Jochen, C. E. (1998). In-line particle characterization. *Part. Part. Syst. Charact., 15*, 47–50.

Scott, D. M., & Gutsche, O. W. (1999). ECT studies of bead fluidization in vertical mills. In T. York (Ed.), *Proceedings of the First World Congress on Industrial Process Tomography.* Buxton, England: VCIPT.

Scott, D. M., & McCann, H., (Eds.). (2005). *Process imaging for automatic control.* Boca Raton, FL: CRC.

Scott, D. M., Sunshine, G., Rosen, L., & Jochen, C. E. (2001). Industrial applications of process imaging and image processing. In H. McCann & D. M. Scott (Eds.), *Process Imaging for Automatic Control.* Proceedings of SPIE 488. Bellingham, WA: SPIE.

Scott, D. M., & Williams, R. A., (Eds.) (1995). *Frontiers in Industrial Process Tomography.* New York: Engineering Foundation.

Sick, V., & McCann, H. (2005). Imaging diagnostics for combustion control. In D. M. Scott and H. McCann (Eds.), *Process imaging for automatic control.* Boca Raton, FL: CRC.

Sunshine, G., Scott, D. M., & Gleich, S. I. (2005). *Method and Apparatus for Measuring Amounts of Non-Cohesive Particles in a Mixture.* U.S. Patent 10/531,922.

Weit, H. (1987). *Betriebsverhalten und Maßstabsvergrößerung von Rührwerkskugelmühlen.* PhD diss., Technical University of Braunschweig.

Williams, R. A., & Beck, M. S., (Eds.) (1995). *Process tomography.* Oxford, England: Butterworth-Heinemann.

Chapter 10

Brekhovskikh, L. M. (1980). *Waves in layered media* (trans. R. T. Beyer). New York: Academic Press.

Currie, I. G. (1974). *Fundamental mechanics of fluids.* New York: McGraw-Hill.

Hartman, P. L. (1987). Interferometer micrometer. *Rev. Sci. Instrum., 58*, 478–479.

Hueter, T. F. & Bolt, R. H. (1955), *Sonics.* New York: Wiley.

Jenkins, F. A., & White, H. E. (1976). *Fundamentals of optics* (4th ed.). New York: McGraw-Hill.

Kogelnik, H. (1979). Propagation of laser beams. In R. R. Shannon & J. C. Wyant (Eds.), *Applied optics and optical engineering* (vol. 7) New York: Academic Press.

Krautkramer, J., & Krautkramer, H. (1983). *Ultrasonic testing of materials* (3rd ed.). New York: Springer-Verlag.

Lefebvre, J. E., Bruneel, C., Delebarre, C. et al. (1988). Remote ultrasonic measurement of the thickness of thin films. *J. Acoust. Soc. Am., 84*, 1094–1096.

Lide, D. R. (Ed.). (1991). *CRC handbook of chemistry and physics* (72nd ed.) Boca Raton, FL: CRC.

Olson, H. F. (1947). *Elements of acoustical engineering* (2nd ed.). New York: Van Nostrand.

Peterson, R. W. et al. (1984). Interferometric measurements of the surface profile of moving samples. *App. Opt., 23*, 1464–1467.

Self, S. A. (1983). Focusing of spherical Gaussian beams. *App. Optics, 22*, 658–661.

Strand, T. C., & Katzir, Y. (1987). Extended unambiguous range interferometry. *App. Opt., 26,* 4274–4281.

Strutt, J. W. (Lord Rayleigh) (1945). *The theory of sound* (1st American ed.), NewYork: Dover.

Xiao, G. Q. et al. (1987). Optical range finder. In D. O. Thompson & D. E. Chimenti (Eds.), *Review of progress in quantitative nondestructive evaluation* (vol. 6A). New York: Plenum.

Chapter 11

Carling, M. J., & Williams, J. G. (1990). Fiber length distribution effects on the fracture of short-fiber composites. *Polymer Composites, 11,* 307–13.

Crandall, E. W., & Jagtap, A. N. (1977). The near-infrared spectra of polymers. *J. App. Pol. Sci., 21,* 449–454.

Filkin, D. L. (1991). *Distributed parallel processing network wherein the connection weights are generated using stiff differential equations.* US Patent 5,046,020.

Hertz, J., Krogh, A., & Palmer, R. G. (1991). *Introduction to neural computation.* Redwood City, CA: Addison-Wesley.

McCulloch, W. S., & Pitts, W. (1943). A logical calculus of the ideas immanent in nervous activity. *Bulletin Math. Biophys., 5,* 115–133.

Palmer, K. F. & Williams, D. (1974). Optical properties of water in the near infrared. *J. Opt. Soc. Am., 64,* 1107–1110.

Quinn, R. A. & Sigl, C. C. (Eds.) (1980). *Radiography in modern industry* (4th ed.). Rochester, NY: Eastman Kodak Company.

Rumelhart, D. E., Hinton, G. E., & Williams, R. J. (1986). Learning representations by back-propagating errors. *Nature, 323,* 533–536.

Scott, D. M. (1989). Density measurements from radioscopic images. *Mater. Eval., 47,* 1113–1119.

Scott, D. M. (1994). *Nondestructive analysis of dispersion and loading of reinforcing material in a composite material.* U.S. Patent 5,341,436.

Scott, D. M. (1995). A two-colour near-infrared sensor for sorting recycled plastic waste. *Meas. Sci. Technol., 6,* 156–159.

Scott, D. M. (1997). Nondestructive analysis of fiber dispersion and loading. *Composites, part A, 28A,* 703–707.

Scott, D. M. (2005). Applications in the chemical industry. In D. M. Scott & H. McCann (Eds.), *Process imaging for automatic control.* Boca Raton, FL: CRC.

Scott, D. M., & Waterland, R. L. (1995). Identification of plastic waste using spectroscopy and neural networks. *Polymer Eng. Sci., 35,* 1011–1015.

Stewart, J. E. (1970). *Infrared spectroscopy.* New York: Marcel Dekker.

Index

D

E

light; See also electromagenetic waves
  incoherent, 44
  monochromatic, 36–37
  polychromatic, 37
  red, 36
  speed, 36
  visible, 36
linear back-projection, 145
lobes, 168–169
log-normal distribution, 136

M

magnetic field, 53
markers, 218–219
matter waves, 19
Maxwell's equations, 35
mean, 9
mean diameter, 100
measurement reproducibility, 11
media mills, 128, 154
  bead fraction, 156–157
melt viscosity, 214
Mie scattering, 109
mill, 155
mirror, 38
model performance, 203
modulators, 42
multilayer films, 184

N

natural sources of radiation, 78
near infrared (NIR) spectrum, 200
negative thermal coefficient (NTC), 80
neural network, 202
  training, 203
neuron activity level, 203
Newton's Law of Gravitation, 49
NIR source, 204
nodal structure, 29
noise; See also error
  sources, 12
noninverting amplifier, 62
normal distribution, 10
normal incidence, 24
normalization
  image, 76, 211
  particle count data, 104
normally distributed data, 11
n-type semiconductor, 58
nucleus, 71

numerical aperture, 179, 182, 193
numerical average, 9

O

observer, 27–28
ohm (Ω), 54
Ohm's Law, 54
ohmic devices, 55
operational amplifier (op-amp), 61
optical cavity, 37, 181
optical elements, 38
optical ranging instrument, 178
optical thickness, 183
optimum grinding, 154
orifice plate, 95

P

parallax, 215–218
partial wave expansion, 115
particle
  counting, 102–108
  diameter definitions, 98
  paste-extruded, 152
  shape, 98, 149–153
  size, 97–98
  size distribution (PSD), 97–101
    algorithm, 137
    cumulative, 100
    differential, 99
pellicle, 40
permittivity of free space, 49
phasors, 58
photodiode, 60
photoelectric effect, 47
photoelectron, 47
photomultiplier tube (PMT), 47
photon correlation spectroscopy, 111
photoresistors, 46
phototransistors, 60
physical observables, 7, 15
pinhole, 41
pixel, 142, 210
Planck's constant, 37, 59
plane wave, 22, 29
plastic recycling, 199
point spread function, 145–146
Poisson counting statistics, 103
polarizability, 57
polarization, 37, 42
polyimide, 183